Barnard Smith, Archibald McMurchy

Elementary Arithmetic for Canadian Schools

Barnard Smith, Archibald McMurchy

Elementary Arithmetic for Canadian Schools

ISBN/EAN: 9783337190811

Printed in Europe, USA, Canada, Australia, Japan

Cover: Foto ©Paul-Georg Meister /pixelio.de

More available books at **www.hansebooks.com**

Howard M. Hoyt's Indestructible Book-Binding, Patented 1879.

THIS BOOK IS BOUND
WITH

HOYT'S PATENT INDESTRUCTIBLE IRON BACK.

THIS remarkable invention entirely does away with the too frequent complaint that books come to pieces in the student's hands, before they have done reasonable service. The *Iron Books* are warranted to remedy this evil. With ordinary care there is practically no limit to their endurance.

DESCRIPTION OF THE IMPROVEMENT

The point of advantage in this binding is simply that, in the opening and closing of the book, there is no friction, which is the destructive element of all ordinary binding.

By the patent method the leaves of the book are fastened by wire pins, inserted near the back margin, the ends of the pins being turned down and lying parallel with the margin, clamping the leaves together.

A back of thin metal is pressed on the book, with flanges overlapping the ends of the pins, which prevents the leaves from tearing from the pins, as is the case with wire, sewing or tape work.

By a peculiar curve in the pins, perfect flexibility is secured in the opening of the volume, the whole forming a practical and substantial binding, absolutely indestructible by ordinary wear and tear.

The Advantage of such a Binding

needs no further demonstration. It fully doubles the value of the volume to which it is applied, and the appearance of a popular series of school books thus invincibly clad will be hailed with enthusiasm by teachers, trustees, parents and pupils. In families the same book will suffice for each child in succession, and School Boards which adopt the FREE BOOK SYSTEM may purchase a single supply which will last for years, passing from hand to hand.

In neatness, convenience, weight and general appearance the IRON BOOKS do not differ appreciably from the ordinary editions. The latter will be continued as long as called for, but it is confidently expected that the Iron Bound edition will rapidly take their place, there being no advance in price, while a general introduction of this improvement will effect at least an

ANNUAL SAVING OF ONE-THIRD.

in the amount at present expended by the people of Canada for school books.

The following will be ready by 1st August, in Iron Binding.

SMITH & McMURCHY'S ELEMENTARY ARITHMETIC.
SECOND BOOK OF READING LESSONS.
THIRD   "   "   "
FOURTH   "   "   "
FIFTH   "   "   "
SPELLING BOOK, *Companion to Readers*
MILLER'S SWINTON'S LANGUAGE LESSONS. (*Revised Edition*).

ADAM MILLER & CO.

SOLE PUBLISHERS OF

Indestructible School Books.

Canadian Series of School Books.

ELEMENTARY ARITHMETIC

FOR

CANADIAN SCHOOLS,

BY

REV. BARNARD SMITH, M.A.,
St. Peter's College, Cambridge;

AND

ARCHIBALD McMURCHY, M.A.,
University College, Toronto.

Authorized by the Council of Public Instruction for Ontario.

TORONTO:
ADAM MILLER & CO., 11 WELLINGTON ST. WEST.
1879.

Entered, according to Act of the Parliament of Canada, in the year one thousand eight hundred and sixty-nine, by the Rev. Egerton Ryerson, L.L.D., Chief Superintendent of Education for Ontario, in the Office of the Minister of Agriculture.

In this ELEMENTARY ARITHMETIC the same method of treating the subject that was adopted in MR. BARNARD SMITH's *Arithmetic for Schools* has been retained; and especial care has been taken to adapt the book, in every respect, to the wants of the Junior Pupils in the Schools of the Dominion.

Presented to

The Library of
The Ontario College of
Education
The University of Toronto

by

Miss Jean Weir

CONTENTS.

SECTION I.

	PAGE
Definitions, Notation and Numeration	9
Notation Table	12
Numeration Table	13
Simple Addition	14
Simple Subtraction	23
Roman Notation	28
Simple Multiplication	ib.
Multiplication Table	29
To Multiply by a number not larger than 12	ib.
To Multiply by a number larger than 12	32
Simple Division	35
To Divide by a number not larger than 12	36
To Divide by a number larger than 12	39

SECTION II.

TABLES. Money	44
Measures of Weight	45
" Length	ib.
" Surface	46
" Solidity	47
" Capacity	48
" Time	49
Reduction	50
Compound Addition	53
" Subtraction	55
" Multiplication	57
" Division	61
To Reduce Old Canadian to the Decimal Currency	65
To Reduce Dollars and Cents to Halifax Currency	ib.
Miscellaneous Examples	66

SECTION III.

Greatest Common Measure	70
Least Common Multiple	71

SECTION IV.

Fractions	73
Vulgar Fractions	74

CONTENTS.

Addition of Vulgar Fractions 8
Subtraction of Vulgar Fractions 8
Multiplication of Vulgar Fractions 8
Division of Vulgar Fractions 8
To Find Value of Vulgar Fractions 88
Reduction of Vulgar Fractions .... ib.
Miscellaneous Examples worked out.................... 89
Decimals 94
To Convert Decimals to Vulgar Fractions 95
Addition of Decimals 96
Subtraction of Decimals 97
Multiplication of Decimals 98
Division of Decimals.................................. 99
To Reduce Vulgar Fractions to Decimals 101
Circulating Decimals................................. 102
Reduction of Decimals 104
Miscellaneous Questions and Examples, Sections I—IV .. 107

SECTION V.

Ratio and Proportion 110
Rule of Three.. 112
Double Rule of Three 118
Practice 123
Interest 127
Simple Interest .... ib.
Compound Interest 129
Present Worth and Discount 131
Present Worth .... ib.
Discount 132
Stocks 134
Per Centage 138
Average 142
Division into Proportional Parts 143
Fellowship or Partnership 144
Simple Fellowship.................................... ib.
Compound Fellowship .... ib.
Equation of Payments 145
Square Root 146
Cube Root 149
Miscellaneous.. 152

SECTION VI.

Mental Arithmetic 155
Answers 160

ARITHMETIC.

SECTION I.

1. ARITHMETIC teaches us the use of NUMBERS.

2. A UNIT or ONE is any single object or thing, as *an orange, a tree.*

3. A WHOLE NUMBER, or AN INTEGER, is a UNIT or ONE, or a collection of UNITS or ONES: if a boy, for instance, have *one* orange, and then another orange is given to him, he will have *two* oranges; if another be given to him he will have *three* oranges; if another, he will have *four* oranges, and so on. One, two, three, four, &c., are called WHOLE NUMBERS or INTEGERS.

4. NOTATION is the art of writing any number in *figures* or *letters.*

There are two methods of Notation: 1st, The ARABIC; 2nd, The ROMAN.

5. The ARABIC NOTATION is the method of expressing numbers by means of the following *figures*, called sometimes *digits.*

   1, 2, 3, 4, 5, 6, 7, 8, 9,
called one, *two, three, four, five, six, seven, eight, nine,* representing (if we express a unit by a dot; thus .),

.	..	...	....	.....	......
or one unit,	or two units,	or three units,	or four units,	or five units,	or six units,

....	....	....
or seven units,	or eight units,	or nine units,

and 0, called *nought*, because when standing by itself, it has no value, and represents nothing. 0 is sometimes called *zero* or *cypher.*

Note. Any one of the figures 1, 2, 3, 4, 5, 6, 7, 8, 9, when standing alone, or as the last figure on the right hand of any number, expresses so many *single* objects or things, or *ones.*

.... or *nine* units are the greatest number of units which can be expressed by one figure.

If another unit be placed to the right hand of the nine units, we have .... or *ten* units, written in figures thus, 10; the 1 in 10 standing in the second place from the right hand, now expresses not *one* unit, but *one ten* units.

Hence we see that although 0, when standing by itself, has no value, still when placed to the right hand of any figure, it *alters* the value of that figure.

The number next after ten represents .... or *eleven* units, written in figures thus, 11, where the 1 in the second place from the right hand expresses *ten* units, and the 1 in the right-hand place of the number *one* unit. Thus 11 units equal 1 ten units and 1 unit more.

Next we come to 12 (*twelve*) or one ten units and 2 more units, 13 (*thirteen*), 14 (*fourteen*), 15 (*fifteen*), 16 (*sixteen*), 17 (*seventeen*), 18 (*eighteen*), 19 (*nineteen*), which represent 1 ten units, and 3, 4, 5, 6, 7, 8, 9, more units respectively.

Next we come to 20 (*twenty*), 21 (*twenty-one*), 22 (*twenty-two*), 23 (*twenty-three*), 24 (*twenty-four*), 25 (*twenty-five*), 26 (*twenty-six*), 27 (*twenty-seven*), 28 (*twenty-eight*), 29 (*twenty-nine*); the 2, when followed by 0 or any single figure, representing *two tens* or *twenty* units, the figures in the right-hand place of each number expressing so many single units.

Next we come to 30 (*thirty*), 31 (*thirty-one*), &c., 3 expressing *three tens*, or *thirty*, and so on up to 40 (*forty*), 4 expressing *four tens* or *forty*, to 50 (*fifty*), to 60 (*sixty*), to 70 (*seventy*), to 80 (*eighty*), to 90 (*ninety*), 5, 6, 7, 8, 9, expressing *five, six, seven, eight, nine tens* respectively: the numbers between any two of them as 40 and 50, being formed in the same way as those between 20 and 30.

We thus come at length to 99 *ninety-nine*) or *nine tens and nine*, the greatest number which can be expressed by two figures.

Ex. I.

Write the following numbers in figures.

(1) Three, four, two, seven, nine, six, eight.

(2) Ten, one, twelve, nineteen, five, eleven, sixteen.

(3) Fourteen, twenty, twenty-seven, thirty-three, forty-nine, sixty, fifty-five, seventeen, thirty-six.

(4) Eighty-eight, thirty-five, sixty-three, twenty-nine, seventy-six, eighty, ninety-four, thirteen, fifty-two.

(5) Write down in figures all the numbers between eight and eighteen, between forty-five and fifty-one, and between eighty-seven and ninety-nine.

The next number after 99 is *one hundred*, written in figures thus, 100; the 1 in 100, standing in the third place from the right hand, now expressing not *one* unit, nor *one* ten units, but *one hundred* units.

All numbers from 100 to 200 (*two hundred*) are formed exactly in the same way, as we formed those from 0 to 100; thus we go on 101 (*one hundred and one*), 102, &c., up to 110 (*one hundred and ten*), then 111 (*one hundred and eleven*), 112, &c., up to 120 (*one hundred and twenty*), then 121, (*one hundred and twenty-one*), 122, &c., up to 130 (*one hundred and thirty*), and so on, up to 200; then 201, 202, &c., up to 300 (*three hundred*), and so on up to 400 (*four hundred*), 500 (*five hundred*), 600 (*six hundred*), 700 (*seven hundred*), 800 (*eight hundred*), 900 (*nine hundred*), 999 (*nine hundred and ninety-nine*), or *nine hundreds, nine tens, and nine*, the greatest number which can be expressed by three figures.

Ex. II.

Write down the following numbers in figures.

(1) One hundred and six, one hundred and fifty, two hundred, two hundred and eighty-seven, three hundred and ten, four hundred and thirty-one, five hundred and fifty-five, nine hundred and nineteen, eight hundred and sixty-seven.

(2) Write all the numbers in figures from one hundred and ninety-five to two hundred and fourteen, from six hundred and eleven, to six hundred and twenty, and from nine hundred and forty-seven to nine hundred and seventy.

The next number after 999 is *one thousand*, written in figures thus, 1000; the 1 in 1000, standing in the fourth place from the right hand, now expressing *one thousand* units.

All numbers from 1000 up to 9999 (*nine thousand nine hundred and ninety-nine*), are formed thus, 1001 (*one thousand and one*), 1002, &c., up to 2000 (*two thousand*), up to 3000 (*three thousand*), and so on.

The next number after 9999 is *ten thousand*, written in figures, thus, 10000; the 1 in 10000 standing in the fifth place

place from the right hand, **now** expressing *one ten thousand units.*

All **numbers from 10000 up to 99999** (*ninety-nine thousand nine hundred and ninety-nine*), are formed thus, 10001 (*ten thousand and one*), 10002, &c., up to 20000 (*twenty thousand*), then 20001 (*twenty thousand and one*), 20002, &c., up to 30000 (*thirty thousand*), and so on.

Ex. III.

Write the following numbers in figures.

(1) Four thousand five hundred and eighty-five, seven thousand three hundred and twenty-one, nine thousand seven hundred and ninety-eight, seven thousand and six.

(2) Five thousand and four, five thousand four hundred, five thousand and forty, eight thousand and thirth-six, eight thousand three hundred and six, eight thousand three hundred and sixty, nine thousand nine hundred and nine.

(3) Seventy-five **thousand six hundred and** thirty-five, ninety thousand **nine hundred and nine, ten** thousand **and four,** eighty-seven **thousand and fifty,** ninety thousand **and one,** sixty-four **thousand and sixty-four,** eighty-three thousand.

The next number after 99999 is *one hundred thousand*, written in figures thus, 100000, the 1 in 100000 standing in the sixth place from the right hand, now expressing *one hundred thousand* units, and so on up to 999999 (*nine hundred and ninety-nine thousand nine hundred and ninety-nine*), then we come to *one million*, written in figures thus, 1000000, the 1 expressing **one** *million* units, and so on up to *tens of millions,* (10000000), *hundreds of millions* (100000000), *billions* (1000000000), **and so on.**

Thus one .... is written	1
ten	10
one hundred	100
one thousand	1,000
ten thousand	10,000
one hundred thousand	100,000
one million	1,000,000
ten million	10,000,000
one hundred million	100,000,000
one billion	1,000,000,000

6. From the above table, **we** see that dividing any num-

ber into periods of three figures each, beginning at the right hand, the names of those periods will be,

First	period	Units.
Second	"	Thousands.
Third	"	Millions.
Fourth	"	Billions.
Fifth	"	Trillions.
&c.	&c.	

Also, that the names of the places in each of those periods are the same, namely:

First	place,	Units.
Second	"	Tens.
Third	"	Hundreds.

7. The following plan is recommended to enable the scholar to write in figures any number dictated by the teacher.

Let the scholar write on his slate a number of noughts, or zeros; thus 000,000,000,000, marking them off into periods of three places each from the right:

Put U over the first period for Units.
T .... second .... Thousands.
M .... third .... Millions.
B .... fourth .... Billions.
    B M T U

And so on. Thus 000,000,000,000. Then when a number is dictated to the pupil, all he has to do is to put each figure under its proper place and fill up vacancies, if any, with 0's.

Thus, two thousand and five will be written thus in figures............................ 2,005
Eighty-six thousand four hundred and three   86,403
Four hundred and thirty thousand three hundred and forty........................... 430,340
Eight hundred and three millions one thousand and eleven...................... 803,001,011
Five billions thirty-seven millions and six .. 5,037,000,006

Ex. IV.

Write the following numbers in figures.

(1) One hundred and five, eight thousand seven hundred and ninety, thirty-seven thousand and seventy-one, thirty thousand four hundred and two, seventy-seven thousand seven hundred, twenty-four thousand eight hundred and seventeen.

(2) One **hundred and five thousand** four hundred and

nine, eight millions eight thousand and thirteen, seven millions six hundred and fifty thousand and ninety, sixty-four millions four hundred, eighty-nine millions forty-four thousand and one, five hundred and four millions six hundred and twenty-three thousand and twenty-four, nine hundred millions three hundred thousand eight hundred, fifty-three millions five hundred and three.

(3) Six billions six millions seventy thousand and seven, eighty-three billions four hundred and one millions one thousand and ten, seven billions and four millions eighty-nine thousand two hundred, nine hundred and ninety millions.

8. NUMERATION is the art of writing in words the meaning of any number, which is already given in figures.

This follows from what has been already said; thus

27 means *two* tens and *seven* units, or twenty-seven.

503 means *five* hundreds, *no* tens, and *three* units, or five hundred and three.

0610 means *no* thousands, *six* hundreds, *one* ten, and *no* units, or six hundred and ten.

5634 means *five* thousands, *six* hundreds, *three* tens, and *four* units, or five thousand six hundred and thirty-four.

6,070,084 means *six* millions, *seven* tens of thousands, *eight* tens and *four* units, or six millions seventy thousand and eighty-four.

803,968,005 means *eight hundreds* of millions, *three* millions, *nine* hundreds of thousands, *six* tens of thousands, *eight* thousands, and *five* units, or eight hundred and three millions nine hundred and sixty-eight thousand and five.

Ex. V.

Write in words the meaning of
- (1) 7, 13, 4, 9, 18, 5, 20, 11, 05, 50, 34, 29, **3, 17, 53.**
- (2) 19, 8, 041, 88, 27, 72, 94, 49, 16, 64, 98, 80, 56, **28.**
- (3) 107, 170, 017, 430, 691, 080, 800, 008, 956, 803, 684.
- (4) 4503, 5870, 5087, 6900, 6099, 02580, 7045, 7591, 6275.
- (5) 24714, 12500, 10025, 10205, 70457, 74007, 77000.
- (6) 300863, 30080630, 96100250, 800400307, 572060495.
- (7) 120192703, 890647560, 1050060429, 100000000001.

SIMPLE ADDITION.

9. SIMPLE ADDITION is the method of finding a number,

SIMPLE ADDITION.

which is equal to two or more numbers of the *same kind* taken together

By the same kind we mean *all* apples, or *all* horses, or *all* pence, and so on.

The numbers to be added are called ADDENDS.

The SUM, or AMOUNT, is the number so found.

Before learning the Rule for Simple Addition, it will be well for a child to learn the following Table, called the ADDITION TABLE. The child should satisfy himself that this Table is true by means of counters, or strokes on a slate.

2 and	3 and	4 and	5 and
1 make 3	1 make 4	1 make 5	1 make 6
2 .. 4	2 .. 5	2 .. 6	2 .. 7
3 .. 5	3 .. 6	3 .. 7	3 .. 8
4 .. 6	4 .. 7	4 .. 8	4 .. 9
5 .. 7	5 .. 8	5 .. 9	5 .. 10
6 .. 8	6 .. 9	6 .. 10	6 .. 11
7 .. 9	7 .. 10	5 .. 11	7 .. 12
8 .. 10	8 .. 11	8 .. 12	8 .. 13
9 .. 11	9 .. 12	9 .. 13	9 .. 14
10 .. 12	10 .. 13	10 .. 14	10 .. 15
6 and	7 and	8 and	9 and
1 make 7	1 make 8	1 make 9	1 make 10
2 .. 8	2 .. 9	2 .. 10	2 .. 11
3 .. 9	3 .. 10	3 .. 11	3 .. 12
4 .. 10	4 .. 11	4 .. 12	4 .. 13
5 .. 11	5 .. 12	5 .. 13	5 .. 14
6 .. 12	6 .. 13	6 .. 14	6 .. 15
7 .. 13	7 .. 14	7 .. 15	7 .. 16
8 .. 14	8 .. 15	8 .. 16	8 .. 17
9 .. 15	9 .. 16	9 .. 17	9 .. 18
10 .. 16	10 .. 17	10 .. 18	10 .. 19

This Table can easily be carried on for numbers larger than 10; for instance since 2 and 1 make 3, 2 and 11 make 10 more than 2 and 1, *i. e.* make 13. Again since 9 and 4 make 13, 9 and 14 will make 23, and so on, the result in each case being 10 more than in the corresponding case in the Table. Also 2 and 51 make 53, 9 and 54 makes 63, and so on, the result in each case being 50 more than the corresponding result in the Table.

10. The sign **+**, called **Plus**, placed between two numbers, means that the numbers are to be added together; thus 2 apples + 3 apples, means that 2 apples and 3 apples are to be added together, therefore 2 apples + 3 apples make 5 apples. Again 2 + 3 + 4 means that 2 and 3 and 4 are to be added together; 2 + 3 make 5, therefore 2 + 3 + 4 make 5 + 4, which make 9.

The sign **=** called **equal**, placed between two numbers, means that the numbers are equal to one another.

The sign ∴ means **therefore**.

Ex. 1. Find the sum of, or add together 5, 4, and 7.

We add thus, 5 and 4 make 9, 9 and 7 make 16;

∴ the sum of 5, 4, and 7, or 5 + 4 + 7 = 16.

```
   5      if so, we say
   4      7 and 4 make
   7      11, 11 and 5
or thus 16  make 16.
```

Ex. 2. Add together 4, 8, 3, 0, 9.

4 and 8 make 12, 12 and 3 make 15, 15 and 0 make 15, 15 and 9 make 24.

∴ 4 + 8 + 3 + 0 + 9 = 24;

```
   4
   8      9 and 3 make 12,
   3      12 and 8 make 20,
   9      20 and 4 make 24.
   0
or thus 24
```

Ex. 3. Find the sum of 9, 3, 7, 6, 5, 9, and 8.

9 and 3 make 12, 12 and 7 make 19, 19 and 6 make 25, 25 and 5 make 30, 30 and 9 make 39, 39 and 8 make 47.

∴ sum of 9, 3, 7, 6, 5, 9, and 8 = 47;

```
   9      8 and 9 make 17,
   3      17 and 5 make 22,
   7      22 and 6 make 28,
   6      28 and 7 make 35,
   5      35 and 3 make 38,
   9      38 and 9 make 47.
   8
or thus 47
```

Ex. VI.

```
Add (1) 2     (2) 3       (3) 5
        3         7           6
        8         8           8
        6         9           7
```

(4), Find the sum of two, seven and two; of five, seven, and four; of six, three, and nine; of five, five, and eight; of nine, eight, nought, and six; of six, two, and nine; of four, eight and three; of seven, nine and two; of nine, five, three, and eight.

SIMPLE ADDITION.

(5). Find the value of $3 + 4 + 8 + 3 + 2 + 5$; $6 + 4 + 0 + 0 + 7 + 2$; $5 + 8 + 1 + 6 + 5 + 9$; $3 + 6 + 8 + 5 + 4 + 2$; $9 + 5 + 7 + 8 + 3 + 4$; $6 + 9 + 9 + 8 + 8 + 5$; $5 + 8 + 3 + 9 + 9 + 6 + 6$.

(6). In a boys' school there are four classes. In the first class there are *six* boys; in the second class *seven* boys; in the third class *one* more than in the first class; in the fourth class *two* more than in the second class. How many boys are there in the school?

(7). John's age is 2 years, Ellen is two years older than John, Walter's age is the sum of the ages of the other two. Find the sum of all their ages.

(8). A woman sold two chickens to A, to B three more than to A, to C as many as to A and B, to D four more than to B; had C bought as many more chickens as he did buy, the woman would have sold all her chickens; how many chickens had she to sell?

Rule for Simple Addition.

11. RULE. Write down the given numbers under each other, so that units may come under units, tens under tens, hundreds under hundreds, and so on: then draw a line under the lowest number.

Find the sum of the column of units: if it be less than ten, write it down under the column of units below the line just drawn, but if it be greater than ten, then write down the units' figure (*i.e.* the last figure on the right hand) of the sum under the column of units, and carry to the column of tens the remaining figure or figures.

Add the column of tens and the figure or figures you carry as you have added the column of units, and treat its sum in exactly the same way as you have treated the column of units.

Treat each succeeding column (viz. hundreds, thousands, &c.) in the same way.

Write down the full sum of the last column on the left hand.

The entire sum thus obtained will be the sum or amount of the given numbers.

Ex. 1. Add together 35, 56, and 282.
By the Rule,

ARITHMETIC.

35
56
282
sum = 373

Method of adding. 2 and 6 are 8, 8 and 5 are 13, *i.e.* 1 ten and 3 units; write down 3 under the column of units, and carry 1 ten.

Then 1 and 8 are 9, 9 and 5 are 14, 14 and 3 are 17, *i.e.* 17 tens, or 10 tens (1 hundred), and 7 tens, write down 7 under the column of tens and carry one hundred.

Then 1 and 2 are 3, *i.e.* 3 hundreds, write down 3 in the hundreds' place.

Ex. 2. Find the sum of three thousand eight hundred and sixty-seven, seven hundred and nine, fifty-six thousand and thirty, eight thousand eight hundred and ninety-six, and fifteen thousand and twenty-nine, and write down the meaning of the sum in words.

By the Rule,

3867
709
56030
8896
15029
84531

eighty-four thousand five hundred and thirty-one.

9 and 6 are 15, 15 and 9 are 24, 24 and 7 are 31, or 3 tens and 1 unit; write down 1 under the units, and carry 3 tens.

Then 3 and 2 are 5, 5 and 9 are 14, 14 and 3 are 17, 17 and 6 are 23, *i.e.* 23 tens, or 2 hundreds and 3 tens; write down 3 tens, and carry 2 hundreds.

Then 2 and 8 are 10, 10 and 7 are 17, **17** and 8 are 25, *i.e.* 25 hundreds, or 2 thousands and 5 hundreds; write down 5 hundreds, and carry 2 thousands.

Then 2 and 5 are 7, 7 and 8 are 15, 15 and 6 are 21, 21 and 3 are 24, *i.e.* 24 thousands, or 2 tens of thousands and 4 thousands; **write down 4 thousands, and** carry 2 tens of thousands.

Then 2 and 1 are 3, 3 and 5 are 8, *i.e.* **8 tens of** thousands; write down 8 tens of thousands.

Note 1. Though the method of adding, as in the above examples, is the one a teacher can follow at first with his pupils; the following method should be insisted on as soon as possible.

Suppose we have to add:

276
389
467
1132

Add thus: 7, 16, 22; put down 2 under the units and 2 to be added to the tens; then 2, 8, 16, 23, &c., &c.: thus saving much time; instead of saying 7 and 9 make 16, 16 and 7 make 23, &c.

Note 2. The truth of all sums in Addition may be proved

SIMPLE ADDITION.

by adding the columns first upwards, and afterwards downwards; if the result be the same in both cases, the numbers will probably have been added correctly.

Ex. VII.

Add	(1) 11 12 14	(2) 22 13 34	(3) 33 45 21	(4) 10 8 81	(5) 27 15 53	(6) 33 22 16	(7) 24 56 35

(8) 12 56 42	(9) 79 27 94	(10) 87 68 59	(11) 98 55 60	(12) 43 69 74	(13) 68 48 98	(14) 78 66 97

(15) 310 46 147	(16) 342 523 876	(17) 704 450 979	(18) 87 867 586	(19) 889 803 509	(20) 500 775 89	(21) 682 962 276

(22) 378 423 748	(23) 797 465 289	(24) 828 939 747	(25) 654 546 465	(26) 729 909 813	(27) 888 517 743	(28) 674 789 555

(29) 2865 758 7632 3403	(30) 785 8756 9540 8559	(31) 6769 8007 5367 7689	(32) 9479 9921 6468 9867	(33) 6045 4500 8068 9647	(34) 5853 9000 8888 5894	(35) 9806 1932 6580 9889

(36) One boy had nineteen marbles, another had seventeen more than the first, and another had nine more than the second, how many marbles had they among them?

(37) In a school section there are two and thirty men, sixty-five more women than men; the number of young men, young women and school children all together equals the number of men and women together, and there are twenty-nine infants; what is the population of the school section?

(38) 5 apple trees produced as follows: the 1st, six hundred and fifty-seven; the 2nd, two hundred and thirty-one

more than the 1st; the 3rd, **eight** hundred and ninety-two; the 4th, eleven more than **all the** first three; the 5th, **as many as** all the others. How **many** apples were there on all the trees?

(39) A gentleman **left his** property by will, thus: **to his wife, nine thousand and eighty** dollars; to each of his two **younger sons, five thousand eight** hundred and ninety-four **dollars; the rest of his property in two** equal shares between his **three daughters and eldest** son; the eldest **son's** share was **fifteen hundred and** twenty dollars more **than the mother's share; what** did the gentleman die **worth?**

(40) A grocer bought 4 chests of oranges. In the **1st** chest there were **five hundred and** eighty-nine oranges; **in the 2nd, two hundred and fifteen more** than in the 1st; in the 3rd, one hundred and **ninety-seven more** than in **the** 1st; in the 4th, as many as **there were in the 1st** and 3rd. How many oranges did he buy?

Ex. VIII.

Add (1) 22 + 30 + 29 + 67   (6) 219 + 315 + 612 + 705
(2) 63 + 93 + 87 + 73   (7) 602 + 528 + 346 + 648
(3) 72 + 90 + 37 + 57 + 39   (8) 736 + 932 + 712 + 836
(4) 38 + 47 + 96 + 83 + 27   (9) 968 + 864 + 345 + 989
(5) 78 + 89 + 68 + 58 + 47   (10) 940 + 760 + 712 + 562

(11)	(12)	(13)	(14)	(15)
71407	82079	96748	33456	15161
90781	88099	25003	84771	8098
68943	67005	84067	66854	958
32600	74387	95674	72984	49790
72777	12345	98765	99999	78368

(16)	(17)	(18)	(19)
9466495	5770821	27591046	768400
7545478	910146	5768000	95320089
29099	6544889	39039587	6949
2988607	7400	596459	84982759
9292929	7683709	78534842	700897
7833210	3684793	19827634	78563412

(20) Add together nine hundred and twelve, two thousand and fifty-eight, three thousand four hundred and forty-five, nineteen thousand three hundred and sixty, twenty-seven thousand six hundred and forty-three, thirty-nine

SIMPLE ADDITION.

thousand seven hundred and ninety, fifty-five thousand eight hundred and seventy-nine, sixty-four thousand nine hundred and seventy-seven, eight thousand two hundred and eleven.

(21) In the census of 1861, the population of the counties on Lake Huron, was as follows: Of Lambton, twenty-four thousand nine hundred and sixteen; of Huron, fifty-one thousand nine hundred and fifty-four; of Bruce, twenty-seven thousand four hundred and ninety-nine; of Grey, thirty-seven thousand seven hundred and fifty; of Simcoe, forty-four thousand seven hundred and twenty. What was the whole population of the above five counties in 1861?

(22) In 1861 the population of the counties on the Ottawa river, was: of Prescott, fifteen thousand four hundred and ninety-nine; of Russell, six thousand eight hundred and twenty four; of Carlton, twenty-nine thousand six hundred and twenty; of Renfrew, twenty thousand three hundred and twenty-five. What was the total population of these four counties in 1861?

(23) In 1861 Toronto contained forty-four thousand eight hundred and twenty-one inhabitants; Montreal, ninety thousand three hundred and twenty-three; Hamilton, nineteen thousand and ninety-six; Ottawa, fourteen thousand six hundred and sixty-nine; Kingston, thirteen thousand seven hundred and forty-three; London, eleven thousand five hundred and fifty-five. Find the total population of these cities in 1861?

Ex. IX.

Find the sum of

(1)	(2)	(3)
20712	2012	22793
212907	75005	27812
616848	700764	38614
703003	93869	45693
1090090	4202573	92075

(4)	(5)	(6)
278653	2612856	37613906
972009	8906783	27305638
2673627	912227	209617382
5009607	6804398	372637867
27693	27635398	968363352
986735	33297653	27306468

ARITHMETIC.

(7)	(8)	(9)
276608567	306738672	397203685
76293568	68345658	28678326
683927285	928327368	206738638
938663589	9283678	728397328
211839297	238906594	563435639
26302562	93567836	912368834
397612397	207867398	6383563
583967323	30673612	83297609
960039368	928327563	603536239
543832586	568302126	736397564
782395678	202386517	932506593

(10)	(11)	(12)
72867853	368936709	378684976
97605812	76385673	79683886
7638516	467308753	468976395
316527308	900009900	786347512
275607836	90909999	927607038
97673904	938568378	90809008
268937318	712050750	758385006
718768926	77807689	703209600
203685738	234593368	87967339
96359568	99213567	862006764
397569387	837346395	993387535

(13) Add together nine millions four hundred and sixty-six thousand four hundred and ninety-five, three hundred and seventy-five millions five hundred and seventy-three thousand seven hundred and thirty-five, seven hundred and fifty-four thousand five hundred and forty-seven, three millions seven hundred and eighty nine thousand two hundred and eighty-four, twenty-nine millions eight hundred and eighty-six thousand seven hundred and ninety-nine, nine hundred and ninety-two thousand and eighty-four, two hundred and ninety-three thousand six hundred and ninety-five, two millions six hundred and eighty-four thousand four hundred and eighty-seven, three millions five hundred and ninety-two thousand eight hundred and seventy-three, seven millions eight hundred and forty-nine thousand three hundred and forty-six.

(14) A farmer had forty-four sheep, thirty-five head of cattle, fifteen pigs, six horses. How many animals had he altogether?

(15) In one year a farmer's crop was as follows: Five hundred and twenty-three bushels of wheat, a hundred and twenty bushels of oats, sixty-four bushels of peas, two hundred and thirty-seven bushels of potatoes, thirty-eight bushels of turnips. How many bushels had he?

(16) A man bought a farm for sixteen hundred and fifty dollars, he spent a hundred and sixty in putting on it new fences, five hundred and seventy-five in building a new house, in repairing the barn and sheds two hundred; he then sold it and made a profit of six hundred dollars. How much did he get for the farm?

(17) In 1861 the population of the counties on Lake Erie was: Essex, twenty-five thousand two hundred and eleven; Elgin, thirty-two thousand and fifty; Kent, thirty-one thousand one hundred and eighty-three; Norfolk, twenty-eight thousand five hundred and ninety; Haldimand, twenty-three thousand seven hundred and eighty; Welland, twenty-four thousand nine hundred and eighty-eight. What was the total population of the six counties on Lake Erie?

SIMPLE SUBTRACTION.

12. SIMPLE SUBTRACTION is the method of finding what number remains, when a smaller number is taken from a greater number of the same kind.

The number so found is called the REMAINDER, or DIFFERENCE.

The number subtracted from, is called the MINUEND; the number subtracted, the SUBTRAHEND.

13. The sign — called MINUS, placed between two numbers, means that the second number is to be subtracted from the first number: thus 7 — 3, or 7 *minus* 3, means that 3 is to be subtracted from 7, ∴ 7 — 3 = 4.

Rule for Simple Subtraction.

14. RULE. Write down the less number under the greater number, so that units may come under units, tens under tens, hundreds under hundreds, and so on; then draw a straight line under the lower number.

Take, if you can, the number of units in each figure of the lower number from the number of units in each figure of the upper number which stands directly over it, and place the remainder under the line just drawn, units under units, tens under tens, and so on.

But, if the units in any figure in the lower number be

greater than the number of units in the figure just above it, then add ten to the upper figure, and then subtract the number of units in the lower figure from the number in the upper figure thus increased, and write down the remainder as before.

Add one to the next number in the lower number, and then take this figure thus increased from the figure just above it, by one of the methods already explained.

Go on thus with all the figures.

The whole difference, or remainder, so written down, will be the difference or remainder of the given numbers.

Ex. 1. Subtract 547 from 859.

By the Rule,

```
      859
      547
diff. = 312
```

Method. 7 from 9 leave 2, *i.e.* 7 units from 9 units leave 2 units; write down 2 in the units' place. 4 from 5 leave 1; *i.e.* 4 tens from 5 tens leave 1 ten; write down 1 in the tens' place.

5 from 8 leave 3, *i.e.* 5 hundreds from 8 hundreds leave 3 hundreds; write down 3 in the hundreds' place.

Ex. 2. Find the difference between seven hundred and forty-two and two hundred and sixty-eight.

By the Rule,

```
      742
      268
diff. = 474
```

I cannot take 8 from 2, *i.e.* 8 units from 2 units, ∴ I add 10 to 2, which makes 12, 8 from 12 leave 4; write 4 in the units' place.

I have added 10 to the upper number 742, I must ∴ add 10 to the lower number 268 (so as not to alter the difference between 742 and 268), *i.e.* 268 must be made 278, or 1 must be added to the 6.

Then I cannot take 7 from 4, *i.e.* 7 tens from 4 tens, ∴ I add 10 to the 4, really 10 tens or 1 hundred to the 4 tens, which makes it 14, really 14 tens, then 7 from 14 leave 7, really 7 tens; write 7 in the tens' place.

I have just added 10 tens, or 1 hundred to the upper number, I must ∴ add 1 hundred to the lower number, *i.e.* I must add 1 to the 2, really 1 hundred to 2 hundreds, making it 3, really 3 hundreds, then 3 from 7 leave 4, really 4 hundreds; write 4 in the hundreds' place.

Ex. 3. How much greater is eight thousand two hundred than six thousand three hundred and nine?

```
      8200
      6309
      ────
      1891
```

9 from 0 I cannot, then 9 from 10 leave 1; write 1 in the units' place; carry 1, really 1 ten, then 1 from 0 I cannot, then 1 from 10 leave 9, really 1 ten from 10 tens leaves 9 tens;

SIMPLE SUBTRACTION.

write 9 in the tens' place; carry 1, really 1 hundred, then 4 from 2 I (cannot, then 4 from 12 leave 8, really 4 hundreds from 12 hundreds leave 8 hundreds; write 8 in the hundreds' place, carry 1, really 1 thousand, then 7 from 8 leave 1, really 7 thousands from 8 thousands leave 1 thousand; write 1 in the thousands' place.

Note. The truth of all sums in subtraction may be proved by adding the less number to the difference or remainder; if this sum equals the larger number, the sum will probably have been worked correctly.

Thus, Proof of Ex. 3. Less number + remainder = 6309 + 1891 = 8200, the greater number.

Ex. X.

	(1)	(2)	(3)	(4)	(5)	(6)	(7)
From	18	27	39	55	86	568	759
Take	14	15	11	5	60	22	603

	(8)	(9)	(10)	(11)	(12)	(13)	(14)
	24	51	64	83	98	70	64
	18	49	6	47	89	54	29

	(15)	(16)	(17)	(18)	(19)	(20)	(21)
	200	547	896	702	800	650	912
	16	380	708	504	199	56	707

	(22)	(23)	(24)	(25)	(26)	(27)	(28)
	563	209	608	486	843	900	505
	476	120	499	307	745	791	107

(29) Subtract thirty-seven from fifty; twenty-nine from seventy-one, sixty-six from one hundred and four; ninety-seven from two hundred and eleven; one hundred and five from three hundred and three, four hundred and seventy-five from six hundred and forty-nine.

(30) A gentleman bought a horse and a carriage for five hundred and sixty dollars, the horse was valued at three hundred dollars. How much was the carriage worth? and how much was the horse worth more than the carriage?

(31) In a school there are 75 children, there are 28 girls. How many more boys than girls are there?

(32) Charles had 167 marbles, he gave John 49, James 65,

Thomas all the rest but 19; how many marbles had Thomas less than James?

(33) By how much does the sum of 6 and 4 exceed their difference?

(34) A boy's father gave him 40 cents to pay 10 cents for a slate, 3 cents for pencils, 8 cents for a copy-book, 5 cents for ink, 3 cents for a postage stamp; after paying for the above he lost all but 4 cents through a hole in his pocket; how much did he lose?

Ex. XI.

	(1)	(2)	(3)	(4)	(5)
From	5467	7601	3000	4536	5480
Take	3546	3890	2001	2297	996

(6)	(7)	(8)	(9)	(10)	(11)
7009	8052	5281	7210	8888	5600
5080	4847	597	3809	999	2575

(12)	(13)	(14)	(15)	(16)
14748	54832	80408	70007	43520
13942	29648	59385	69999	25347

(17)	(18)	(19)	(20)
445673	9200000	87125391	650030042
277594	560506	68050092	94090096

(21) What number taken from three thousand will leave one hundred and one? What number added to seventy-two thousand five hundred and seventy-six will make one million seventy thousand four hundred and nine.

(22) The sum of three numbers is twenty-three thousand two hundred and fifty-seven; the first is 9277, and the second is twelve hundred and eighty-three less than the first; find the third number.

(23) What is the difference between 23047 + 175 − 368 + 495 − 132 and 10000 − 8406 − 704 + 7305?

(24) When will the Prince of Wales, who was born in the year 1841, be as old as the Queen now, in the year 1869, is, who was born in the year 1819? How old will the Queen then be?

(25) John says to Henry, I have 97 marbles; Henry re-

plies, I have 29 less than you; Charlie adds, I have as many as both of you, less 25. How many marbles had Henry, and how many had Charlie?

(26) A man whose yearly income is 1000 dollars, spends 84 dollars for house rent, 135 dollars for servants, 39 dollars in travelling, 58 dollars in clothing, as much on his garden as in travelling and clothing, 804 dollars in household bills. Will he have saved anything, or be in debt at the end of the year, and to what amount?

(27) Harry goes up sixteen steps of a ladder, which has 45 steps, then down 7 steps, then up 10, then down 2, then down 4, then up 11, then down 9, then up 7, then up 5, then down 8, what step from the top and bottom will he then be standing upon?

(28) In a union workhouse there are 133 inmates. The number is made up thus: infirm and able-bodied 70; able-bodied and children 105; children and officers 63; officers 5. Find the number of each class.

(29) A basket contained oranges, nuts, and eggs; in all 1769; there were 1696 oranges and nuts, and 1262 nuts and eggs. How many more nuts were there than oranges?

(30) The population of the counties on the river St. Lawrence in 1861, was one hundred and seventeen thousand nine hundred and eighty-six, that of those on the Ottawa river was seventy-two thousand two hundred and sixty-eight. Find the difference between the population of these counties?

(31) What is the difference between thirty-seven millions nine hundred and six thousand seven hundred and three, and forty-five millions three thousand and eight?

(32) The subtrahend is fifty-six millions two hundred and twelve thousand three hundred, the remainder seventy-seven thousand three hundred and thirteen. What is the minuend?

(33) The minuend is sixty-six millions three hundred and four thousand, the difference twelve thousand five hundred and eighty-six. Find the subtrahend.

(34) A man bought 305 sheep for 3 dollars a head, and after spending 45 dollars on them for food, sold them for 4 dollars a head; how many dollars did he gain by his bargain?

(35) For the year 1861 the Imports into Canada were forty-three millions fifty-four thousand eight hundred and thirty-six dollars, and the Exports were thirty-four millions

seven hundred and seventeen thousand two hundred and forty-eight dollars. Find by how much the Imports exceeded the Exports for the year 1861.

15. ROMAN NOTATION. I, denotes one; V, five; X, ten; L, fifty; C, one hundred; D, five hundred; M, one thousand.

RULE. Where any one of the above letters is *after*, or to the right hand of, one of equal or greater value, it is to be *added* to it, but when put *before* one of greater value, it is to be *subtracted* from it.

Thus II = 1 + 1 = 2, III = 1 + 1 + 1 = 3, IV = 5 less 1 = 4, VI = 5 + 1 = 6, VIII = 5 + 1 + 1 + 1 = 8, IX = 10 less 1 = 9, XIII = 10 + 1 + 1 + 1 = 13, XIV = 10 plus 5 less 1 = 10 + 4 = 14, LXXIX = 50 + 10 + 10 + 10 less 1 = 70 + 9 = 79, XC = 100 less 10 = 90.

Note. A line over a letter or letters, increases their value a thousandfold: thus V=5, V̄=5000; C=100, C̄=100000.

Ex. XII.

1. Express in the Roman Notation, three; seven; eleven; nine; twelve; sixteen; 18; 25; 28; 37; 40; 53; 59; 62; 77; 84; 103; 157; 190; 200; 651; 783; 1204; 1527; 1865.

2. Express in words, and also in Arabic figures, III; VI; VIII; XIII; XV; XVII; XX; LIV; LXXXI; CXI; DCV; V̄II; M̄C̄; MM; DCCXLIX; MDCCCLXV.

SIMPLE MULTIPLICATION.

16. SIMPLE MULTIPLICATION is a short method of repeated addition; thus, when 2 is multiplied by 3, the number obtained is the sum of 2 repeated three times, which sum = 2 + 2 + 2 = 6.

The number, which is to be repeated or added to itself, is called the MULTIPLICAND; thus, in the above example, 2 is the multiplicand.

The number which shows how often the multiplicand is to be repeated, is called the MULTIPLIER; thus, in the above example, 3 is the multiplier.

The number found by multiplication, for instance 6 in the above example, is called the PRODUCT.

The multiplier and multiplicand are sometimes called FACTORS, because they are factors, or makers, of the product.

The sign ×, called INTO, OR MULTIPLIED BY, placed be-

SIMPLE MULTIPLICATION.

tween two numbers means that the numbers are to be multiplied together.

The following Table, called the MULTIPLICATION TABLE, ought to be learned correctly:

Twice 1 makes 2	3 times 1 makes 3	4 times 1 makes 4	5 times 1 makes 5	6 times 1 makes 6	7 times 1 makes 7
2 .. 4	2 .. 6	2 .. 8	2 .. 10	2 .. 12	2 .. 14
3 .. 6	3 .. 9	3 .. 12	3 .. 15	3 .. 18	3 .. 21
4 .. 8	4 .. 12	4 .. 16	4 .. 20	4 .. 24	4 .. 28
5 .. 10	5 .. 15	5 .. 20	5 .. 25	5 .. 30	5 .. 35
6 .. 12	6 .. 18	6 .. 24	6 .. 30	6 .. 36	6 .. 42
7 .. 14	7 .. 21	7 .. 28	7 .. 35	7 .. 42	7 .. 49
8 .. 16	8 .. 24	8 .. 32	8 .. 40	8 .. 48	8 .. 56
9 .. 18	9 .. 27	9 .. 36	9 .. 45	9 .. 54	9 .. 63
10 .. 20	10 .. 30	10 .. 40	10 .. 50	10 .. 60	10 .. 70
11 .. 22	11 .. 33	11 .. 44	11 .. 55	11 .. 66	11 .. 77
12 .. 24	12 .. 36	12 .. 48	12 .. 60	12 .. 72	12 .. 84

8 times 1 makes 8	9 times 1 makes 9	10 times 1 makes 10	11 times 1 makes 11	12 times 1 makes 12
2 .. 16	2 .. 18	2 .. 20	2 .. 22	2 .. 24
3 .. 24	3 .. 27	3 .. 30	3 .. 33	3 .. 36
4 .. 32	4 .. 36	4 .. 40	4 .. 44	4 .. 48
5 .. 40	5 .. 45	5 .. 50	5 .. 55	5 .. 60
6 .. 48	6 .. 54	6 .. 60	6 .. 66	6 .. 72
7 .. 56	7 .. 63	7 .. 70	7 .. 77	7 .. 84
8 .. 64	8 .. 72	8 .. 80	8 .. 88	8 .. 96
9 .. 72	9 .. 81	9 .. 90	9 .. 99	9 .. 108
10 .. 80	10 .. 90	10 .. 100	10 .. 110	10 .. 120
11 .. 88	11 .. 99	11 .. 110	11 .. 121	11 .. 132
12 .. 96	12 .. 108	12 .. 120	12 .. 132	12 .. 144

17. *Rule for Simple Multiplication, when the multiplier is a number not larger than 12.*

RULE. Place the multiplier under the multiplicand, units under units, and (if the multiplier be 10, 11, or 12) tens under tens; then draw a line under the multiplier.

Multiply each figure of the multiplicand, beginning with the units, by the figure, or figures of the multiplier (by means of the Multiplication Table).

Write down and carry as in Simple Addition.

30 ARITHMETIC.

Ex. 1. Multiply **531 by 2.**
By the Rule.

```
 531
  ·2
————
1062
```

Twice **1** unit **makes 2** units; write **2 in the** units' place of the product. Twice 3 tens of units make 6 tens of units; write 6 in the tens' place of the product. Twice 5 hundreds of units make 10 hundreds of units, or 1 thousand 0 hundred; write 0 in the hundreds' place, and 1 in the thousands' place.

Ex. 2. Find the product of 5063 and **6.**
By the Rule.

```
 5063
    6
—————
30378
```

6 times 3 units = 18 units = 1 ten and 8 units; write 8 units, carry 1 ten. Next 6 times 6 tens = 36 tens, which added to the 1 ten carried = 37 tens = 3 hundreds and 7 tens; write 7 tens and carry 3 hundreds.

Next 6 times 0 hundreds = 0, which added to the 3 hundreds carried = 300 hundreds, write 3 in the hundreds' place.

Next, **6 times 5** thousands = 30 thousands = 3 tens of thousands and 0 thousands; write 0 in the thousands' place, and 3 in the tens of thousands' place.

Note. It will be seen from the Multiplication Table, that to multiply any **number** by ten, we have only to write 0 to the right-hand of the number, thus, $3 \times 1 = 3$, $3 \times 10 = 30$; also $5893 \times 10 = 58930$, and $58930 \times 10 = 589300$.

Similarly $3 \times 100 = 300$, $3 \times 1000 = 3000$, and so on.

Also if any **number be multiplied** by 20, the result is the same as if the **number were** multiplied by 2, and 0 written on the right hand of the product; thus, $6 \times 20 = 6 \times 2 \times 10 = 12 \times 10 = 120$; also, $60 \times 20 = 1200$, for $60 \times 20 = 60 \times 2 \times 10 = 120 \times 10 = 1200$; and so of any other number.

Similarly $60 \times 200 = 12000$, $60 \times 2000 = 120000$, and so on.

Ex. XIII.

	(1)	(2)	(3)	(4)	(5)	(6)	(7)	(8)	(9)
Multiply	53	47	88	56	48	60	29	75	27
By	2	2	2	2	3	3	3	3	4

(10)	(11)	(12)	(13)	(14)	(15)	(16)	(17)	(18)
51	83	90	67	43	36	99	78	27
4	4	5	5	5	6	6	6	7

SIMPLE MULTIPLICATION.

(19)	(20)	(21)	(22)	(23)	(24)	(25)	(26)	(27)
53	45	77	69	54	20	99	53	87
7	8	8	9	9	10	10	11	11

(28)	(29)	(30)	(31)	(32)	(33)	(34)	(35)	(36)
91	60	49	687	800	697	276	777	497
11	12	12	2	3	3	4	5	6

(37)	(38)	(39)	(40)	(41)	(42)	(43)	(44)
479	905	835	487	560	538	888	704
7	7	8	9	10	11	12	12

(45) Supposing an acre of land to produce 39 bushels of wheat, how many bushels will 11 of such acres produce, and what will be their value at 6 shillings a bushel?

(46) There are 21 shillings in 1 guinea, and 12 pence in 1 shilling; how many pence are there in 3, 7, 12 guineas?

(47) Charlie bought of Quintin 11 rabbits at 23 cents each, and Quintin bought of Charlie 9 hens at 33 cents each, how many cents had Quintin to give to Charlie?

(48) What is the difference between 12 dozen and 8, and 8 dozen and 12? [*Note*, 1 dozen = 12.]

(49) *A* has seven thousand four hundred and one potatoes; he sells *B* fifty-seven dozen and five; *C* one hundred and twelve dozen and eleven; *D* two hundred and fifty-nine dozen and nine; and *E* the remainder. How many more did *E* buy than *C*?

Ex. XIV.

	(1)	(2)	(3)	(4)	(5)	(6)
Multiply	9048	5849	9873	38076	6057	97068
By	2	2	3	3	4	5

(7)	(8)	(9)	(10)	(11)	(12)
69360	80965	43909	48508	33069	38476
6	5	7	8	7	9

(13)	(14)	(15)	(16)
49216	69432	21357	91537
12	12	11	12

ARITHMETIC.

(17) Multiply (1) 3870492, (2) 4609758, (3) 85973864, (4) 9090853, (5) 55880092, (6) 987654321, by each of the following, 2, 5, 3, 7, 4, 9, 6, 8, 11, and 12.

(18) Two persons start from the same place, and travel in the same direction, one at the rate of 93 miles a day, the other at the rate of 79 miles a day; how far apart will they be at the end of a week?

(19) If the second person at the end of two days turn back, and travel each day in the opposite direction the same number of miles as before; how far will they be apart at the end of a week?

18. *Rule for* **Simple Multiplication,** *when the Multiplier is a number larger than 12.*

RULE. Place the multiplier under the multiplicand, units under units, tens under tens, and so on; then draw a line under the multiplier.

Multiply each figure of the multiplicand, beginning with the units, by the figure in the units' place of the multiplier (by means of the table given for Multiplication); write down and carry as in Addition.

Then multiply each figure of the multiplicand, beginning with the units, by the figure in the tens' place of the multiplier, placing the first figure so obtained under the tens of the line above, the next figure under the hundreds, and so on.

Proceed in the same way with each succeeding figure of the multiplier.

Then add up all the results thus obtained by the **rule of** Simple Addition.

Ex. Multiply 2307 by 358.

By the Rule,

```
         2307
          358
        ─────
        18456
        11535
         6921
        ─────
product = 825906
```

since $358 = 300 + 50 + 8$, when we multiply by the 5, we in fact multiply by 50, and $2307 \times 50 = 115350$; again, when we multiply by the 3, we in fact multiply by 300, and $2307 \times 300 = 692100$; hence it is quite clear that we may multiply by the simple figures 5 and 3, if we only take care to place the first figure in the second line under the tens' place of the first line, and the first figure of the third line under the hundreds' place.

SIMPLE MULTIPLICATION.

Ex. 2. Find the product of 758 and 609.

```
    758
    609
  ─────
   6822
   4548
  ─────
 461622
```

Since 758, or any other number, multiplied by 0 gives 0 as a product, ∴ in this case we multiply by 9 and then by 6, writing the first figure of the second line under the hundreds' place, and not under the tens' place of the line above, for $609 = 600 + 9$.

Note 1. If the MULTIPLIER or MULTIPLICAND, or both, end with cyphers, we may omit them in the working; taking care to place on the right hand of the product as many cyphers as we have omitted from the end of the multiplier or multiplicand, or both. Thus, if 270 be multiplied by 507, and 2700 be multiplied by 50700, we have

```
    270            270
    507          50700
   ────          ─────
    189            189
    135            135
   ─────         ───────
  136890         1368900
```

In the first case, when we multiply 7 by 7, in fact we multiply 70 by 7, and $70 \times 7 = 490$.

In the second case, when we multiply 7 by 7, in fact we multiply 70 by 700, and $70 \times 700 = 49000$.

Note 2. $2 \times 3 = 2 + 2 + 2 = 6$, and $3 \times 2 = 3 + 3 = 6$
∴ $2 \times 3 = 3 \times 2$; and this is true of all numbers.

Note 3. If more than two factors have to be multiplied together, as $2 \times 4 \times 9$, it is termed CONTINUED MULTIPLICATION, and since $2 \times 4 = 8$, and $8 \times 9 = 72$, and ∴ $2 \times 4 \times 9 = 72$, we shall of course obtain the same result, whether we multiply any number by 72, or by its factors 2, 4, and 9, by continued multiplication, and so of any other number.

$35 \times 72 = 2520$, and $35 \times 2 \times 4 \times 9 = 70 \times 4 \times 9 = 280 \times 9 = 2520$.

19. Numbers which are produced by multiplying together two or more numbers respectively greater than unity, are called COMPOSITE NUMBERS. Thus $4 = 2 \times 2$, $36 = 6 \times 6$, or $= 2 \times 3 \times 2 \times 3$, and such like, are COMPOSITE NUMBERS.

Numbers which cannot be broken up into factors, as 3, 5, 7, 11, and such like, are PRIME NUMBERS.

Note 4. The truth of all results in Multiplication may be proved by using the multiplicand as multiplier, and the multiplier as multiplicand; if the product thus obtained be the same as the product found at first, the results are in all probability true.

Ex. XV.

	(1)	(2)	(3)	(4)	(5)	(6)
Multiply	463	678	276	601	946	837
By	18	27	33	54	61	39

(7)	(8)	(9)	(10)	(11)	(12)	(13)	(14)
793	407	869	917	692	909	305	463
30	55	89	46	73	88	715	608

(15)	(16)	(17)	(18)
1263	5613	96732	67628
36	54	72	64

(19)	(20)	(21)	(22)	(23)	(24)	(25)
495	690	417	278	904	3259	15900
370	480	739	900	803	497	3300

(26)	(27)	(28)	(29)	(30)	(31)
50738	86370	47672	68109	45094	56888
9706	90900	5126	2065	7838	6049

(32)	(33)	(34)	(35)	(36)
92035	84009	678000	90058	80108
8007	7898	876000	90009	7770

(37) Find the product of seven thousand and thirty-nine by four thousand seven hundred and nine; three thousand nine hundred and ten by three hundred and fifty thousand; eighty-seven thousand nine hundred by nine thousand and six; seven millions eight thousand and five by four hundred thousand seven hundred and three.

(38) Find the product of the sum and difference of four hundred and ninety-six, and three hundred and twelve.

(39) Multiply (1) 973 by 63, and also by its factors 3, 3, and 7, and (2) 33000 by 1560, and also by its factors 13, 5, 4, and 6.

(40) As in (39) do also, (15), (16), (17), (18).

Ex. XVI.

	(1)	(2)	(3)	(4)
Multiply	78689	275832	729817	46481
By	547	476	6736	936

SIMPLE DIVISION.

```
   (5)        (6)        (7)        (8)         (9)
 40930      9264397    6707936    6078908    708670567
   779         9584       9878       6725        97806
```

```
    (10)         (11)         (12)
  6835675      27083679     25058612
    2689          3709         6289
```

(13) Find the product of 3523725 and 2538.
(14)   "   "   2778588 and 9867.
(15)   "   "   79068025 and 1386.
(16)   "   "   79094451 and 764095.

(17) Multiply five millions seventy-six thousand eight hundred and twelve by ninety-seven thousand six hundred and thirteen.

(18) Multiply nine millions five hundred and seven thousand three hundred and forty by seven thousand and seventy-one.

(19) Required the product of twelve millions four hundred and eighty-one thousand six hundred and thirty, and fifteen hundred and nine.

SIMPLE DIVISION.

20. SIMPLE DIVISION is a short method of repeated Subtraction; or, it is the method of finding how often one number called the DIVISOR is contained in another number called the DIVIDEND. The number, which shews this, is called the QUOTIENT.

Thus, the dividend 12 divided by the divisor 4 gives the quotient 3; and for this reason, $4 + 4 + 4 = 12$, and therefore if we subtract 4 from 12, and then a second 4 from the remainder 8, and then a third 4 from the remainder 4, nothing remains.

If however some number be left, after the divisor has been taken as often as possible from the dividend, that number is called the REMAINDER; thus, 11 divided by 4 gives a quotient 2, and a remainder 3; for after subtracting 4 from 11 once, there is a remainder 7; after subtracting 4 a second time from the remainder 7, there is a remainder 3.

The sign $\div$, called BY, or DIVIDED BY, placed between

two numbers, signifies that the first is to be divided by the second.

Division is just the opposite of Multiplication. By the Multiplication Table, $3 \times 4 = 12$, and $12 \div 4 = 3$, or $12 \div 3 = 4$.

21. *Rule for* **Simple Division,** *when the Divisor is a number not larger than 12.*

RULE. Place the divisor and dividend thus:
divisor) dividend.

Take off from the left hand of the dividend the least number of figures which make a number not less than the divisor.

Find by the Multiplication Table how often the divisor is contained in this number; write the quotient under the units' figure of this number, and take notice of the remainder, whether it be any number or 0.

On the right of the remainder (whether it be any number or 0), conceive in your mind to be placed the least number of the figures next following in the dividend which will, affixed to the remainder, make a number not less than the divisor. Proceed, as above, with this new dividend to find the next figure of the quotient; taking care to place after the first figure in the quotient a cypher for every figure just brought down from the dividend except the last.

Continue this process till all the figures in the dividend have thus been brought down.

If there be a remainder at the end of the operation, write it as a remainder distinct from the quotient.

Ex. 1. Divide 756 by 3.
By the Rule,
3)756
  252

Method of working. 3 in 7 goes 2 times and 1 over, write 2 under the 7; 3 in 15 goes 5 times, write 5 under the 5; 3 in 6 goes 2 times, write 2 under the 6.

Reason. In 756 the $7 = 700$, the 5 50, $=$ and the $6 = 6$. Now 3 in 700 goes 200 times, and 100 over, therefore write 2 in the hundreds' place, and carry the 100; then 3 in 100 + 50, or 150, goes 50 times, therefore write 5 in the tens' place; then 3 in 6 goes 2 times, therefore write 2 in the units' place.

SIMPLE DIVISION

Ex. 2. Find the quotient of 21406 by 7.

7)21406
    ‾‾‾‾‾
    3058

Method of working. 7 in 2 goes no times, but 7 in 21 goes 3 times, write 3 under the 1; 7 in 4 goes no times, 7 in 40 goes 5 times and 5 over; write 0 under the 4 and 5 under the 0; then 7 in 56 goes eight times, write eight under the 6.

Reason. In 21406 the 21 is = 21000, the 4 = 400, and the 6 = 6.

∴ the 3 in the quotient = 3000, the 5 = 50, the 8 = 8. and the quotient is 3058.

Ex. 3. Into how many classes of eleven each can a population of eight hundred and ninety thousand three hundred and eighty-nine to be divided?

11)890389
   ‾‾‾‾‾‾
   80944 rem. 5.

i. e. 80944 classes and 5 people over, or
890389 = 80944 × 11 + 5

11 in 8 will not go, 11 in 89 goes 8 and 1 over, write 8 under the 9; 11 in 10 will not go, 11 in 103 goes 9 and 4 over, write 0 under the 0, and 9 under the 3; 11 in 48 goes 4 and 4 over, write 4 under the 8; 11 in 49 goes 4 and 5 over, write 4 under the 9, and rem. 5.

Ex. 4. Distribute six hundred thousand four hundred and fifty-five apples in equal portions between 12 families.

12)600455
   ‾‾‾‾‾‾
   50037 rem. 11

∴ each family receives 50037 apples, and there are 11 apples over; or
600455 less 11 = 50037 × 12

12 in 60 goes 5; for the next dividend we have 045, ∴ we write two cyphers or 00 after the 5; 12 in 45 goes 3 and 9 over, ∴ write 3 after 0; then 12 in 95 goes 7 and 11 over, ∴ write 7 after 3, and rem. 11.

Ex. XVII.

Note. Each of the given **numbers is to be divided by each** of the different divisors.

(1) 88, 93, 98, 103, 100, by 6, 9, and 8.
(2) 105, 110, 119, 128, 117, by 5, 11 and 10.
(3) 130, 141, 153, 168, 147, by 6, 12, and 11.
(4) 172, 195, 206, 257, 240, by 6, 8, and 12.
(5) 462, 682, 840, 405, 555, by 4, 10, and 11.
(6) 600, 763, 842, 999, 717, by 11, 8 and 12.
(7) 1210, 6876, 6063, 5000, by 9, 12, and 11.

ARITHMETIC

(8) 2760, 9604, 8267, 6548, by 8, 12, and 10.

(9) 86246, 72635, 85490, 35298, by 12, 10, and 7.

(10) 76002, 90009, 53027, by 11, 8, and 12.

(11) 5470698, 93700682, 2060198, by 8, 10, and 11.

(12) 8360047, 6789643, 9889989, by 7, 9, and 12.

(13) How many times can you subtract twelve from eight hundred thousand seven hundred and nine? What number besides 11 will exactly divide 218581?

(14) (1) If the dividend be 84, the quotient 9, the remainder 3, what is the divisor? (2) If the divisor be 11, the remainder 7, the quotient 146, what is the dividend?

(15) A woman bought 11 fowls at 36 cents each, and sold them so as to gain 198 cents; what did she sell each fowl for?

(16) A boy, having a basket containing 214 plums, distributed them equally between his eight schoolfellows and himself; the number which remained over he gave to his schoolmaster; how many did the schoolmaster receive?

(17) The sum of two numbers is 4563, and the less number is 9; find their quotient.

(18) Find the difference between the product of 40687 and 503, and the quotient of 93710562 by 11.

(19) A Bachelor, who died worth 5427 dollars, left 1500 dollars to charities, and the rest of his property between his housekeeper, manservant, and cook; the manservant was to have twice the cook's share, and the housekeeper was to have twice the manservant's share; what did each receive.

(20) If the sum of 18 and 30 be divided by their difference, and the quotient be multiplied by the product of 16 and 27, what is the result?

(21) Find the product of nine hundred and seven thousand and fifty seven by six millions and six, and find what number added to the result will make it exactly divisible by nine.

(22) A basket contained 282 apples and oranges; there were 230 more apples than oranges. Find the number of oranges.

(23) How many penknives, worth 16 cents each, ought to be exchanged for 4 gross of penholders at 10 cents per dozen, and twenty-five score envelopes at 16 cents a hundred? *Note*, 1 score = 20, 1 gross = 12 dozens.

22. *Rule for Simple Division, when the Divisor is a number larger than 12.*

RULE. Place the divisor and dividend thus:

divisor) dividend (

leaving a space for the quotient on the right of the dividend.

Take off from the left hand of the dividend the least number of figures which make a number not less than the divisor.

Find how many times the divisor is contained in this number; write the quotient as the left-hand figure of the whole quotient; multiply the divisor by this figure, and bring down the product under the number taken off from the left of the dividend, and subtract.

On the right of the remainder (whether it be any number or 0) place the least number of figures next following in the dividend which will, affixed to the remainder, make a number not less than the divisor. Proceed as above with this new dividend to find the next figure of the quotient; taking care to place after the first figure in the quotient a cypher for every figure just brought down from the dividend except the last.

Continue this process till all the figures in the dividend have thus been brought down.

If there be a remainder at the end of the operation, write it as a remainder distinct from the quotient.

Note. If any remainder be equal to or greater than the divisor, the last figure of the quotient must be changed for one greater.

Ex. 1. Divide 1368 by 57.

By the Rule,

57) 1368 (24
114
———
228
228
———

Method of Working. 136 is the *least* number taken from the left of the dividend, into which 57 will go; we then say 5 into 11 goes 2; write 2 as the first figure of the quotient on the right hand, write also 114 (product of 57 × 2) under 136 and subtract; we obtain a remainder 22. Then place 8, the next figure in the dividend, to the right of the remainder; we thus obtain a new dividend 228; as before 5 into 22 goes 4; write the 4 to the right of the 2 in the quotient; and so proceed till all the figures in the dividend are brought down.

Reason. $1368 = 1360 + 8$; $\therefore$ the 1st dividend is really 1360; now $57 \times 20 = 1140$, $\therefore$ the 1st number in the quotient is 20; and $1360 - 1140 = 220$; $\therefore$ the second dividend is $220 +$

40   ARITHMETIC.

8 or 228, and as $57 \times 4 = 228$, ∴ the second figure in the quotient is 4, and the quotient is $20 + 4$ or 24.

Note. Since $1368 \div 57 = 24$, it follows that $1368 \div 24 = 57$, and also that $57 \times 24 = 1368$.

Ex. 2. Find the quotient of 1039888 by 5048.

```
5048 ) 1039888 ( 206
       10096
       -----
        30288
        30288
```
10398 is the *least* number, taken from the left of the dividend, into which 5048 will go; we then say 5 in 10 goes 2, and $5048 \times 2 = 10096$; write 2 as the left-hand figure of the quotient, 10096 under 10398, and subtract; we obtain a remainder 302. Then we have to place the next *two* figures 88 of the dividend to the right of this remainder to form a number 30288 greater than the divisor, ∴ we must write 0 in the quotient after 2; then 5 in 30 goes 6 times, and $5048 \times 6 = 30288$, write 6 in the quotient after 0, 30288 under 30288, and subtract; there being no remainder, 206 is the quotient required.

Ex. 3. How many times does 318493585 contain 8607 ?

```
8607 )318493585( 37004
      25821
      -----
       60283
       60249
       -----
        34585
        34428
        -----
          157
```
After obtaining 37 in the quotient, 3 figures of the dividend have to be brought down to get the next significant figure in the quotient, ∴ write *two* cyphers in the quotient.

8607 is contained 37004 times in 318493585, and there is a remainder 157; in other words $318493585 = 37004 \times 8607 + 157$, or 318493585 less $157 = 37004 \times 8607$.

23. When the divisor is a composite number, and made up of two factors, neither of which exceeds 12, the dividend may be divided by one of the factors in the way of Short Division, and then the result by the other factor. If there be a remainder after each of these divisions, the true remainder will be found by multiplying the second remainder by the first divisor, and adding to the product the first remainder.

Ex. 4. Divide 56732 by 45.

```
45 { 9 | 56732, i. e. 56732 units,
     5 | 6303 rem. 5, i. e. 6303 nines and rem. 5 units,
         1260 rem. 3, i.e. 1260 forty-fives, and rem. 3 nines,
```

∴ the **true rem.** $= 9 \times 3$ units $+ 5$ units $= 27 + 5$, or 32 units.

SIMPLE DIVISION.

Therefore the quotient arising from the division of 56732 by 45 is 1260 with a remainder 32 over.

Ex. XVIII.

Divide

(1) 192 by 16 ; 720 by 18 ; 795 by 15 ; 1786 by 19.
(2) 1035 by 23 ; 1073 by 37 ; 2730 by 42 ; 5432 by 56.
(3) 4560 by 80 ; 3871 by 49 ; 7744 by 88 ; 6935 by 95.
(4) 5375 by 25 ; 29526 by 37 ; 25665 by 29 ; 4590 by 45.
(5) 69230 by 86 ; 37510 by 55 ; 10287 by 81 ; 23919 by 67 ; 25760 by 56 ; 538840 by 76.
(6) 35626 by 94 ; 31339 by 77 ; 80840 by 86 ; 28782 by 39 ; 9009196416 by 96 ; 41765256 by 72.
(7) 88832 by 256 ; 175252 by 308 ; 321776 by 104.
(8) 653723 by 329 ; 3577926 by 506 ; 542100 by 834.
(9) 8189181 by 909 ; 4049820 by 745 ; 342604 by 883.
(10) 7848600 by 365 ; 2339100 by 678 ; 90625 by 727.
(11) 2729188 by 478 ; 30387310 by 397 ; 3273068 by 703.
(12) 37624792 by 843 ; 90273189 by 513 ; 53006751 by 609 ; 30073074 by 338 ; 6307625409 by 652.
(13) 519387042 by 2731 ; 10101255 by 2185 ; 154725876 by 3076 ; 632795014 by 7243.
(14) 2015029 by 1004 ; 131686100 by 6487 ; 395194875 by 6007 ; 506961584 by 1617.
(15) 4519559744 by 5008 ; 16322853 by 9306 ; 23617103000 by 1579 ; 2106144185 by 2735.
(16) 142997420 by 3782 ; 19554707200 by 6016 ; 282888270157S by 38706.

(17) What number multiplied by 79 will give the same product as 257 multiplied by 553 ?

(18) How many pairs of stockings, at 60 cents a pair, should be given for 9 dozen pairs of gloves, at 110 cents a pair ?

(19) What number must be added to thirty millions nine hundred and eighty-four thousand and fifty-one, that the sum may be exactly divisible by two hundred and eighty-eight ?

(20) **If** the sum of 274 and 108 be multiplied by their

difference, and the product be divided by 176, what will be the quotient?

(21) A farmer bought 75 sheep at 4 dollars each; 94 sheep at 3 dollars each; and 106 sheep at 2 dollars each; at what price per head must he sell the sheep, so as to gain 147 dollars by his bargain?

(22) A hatter sold 267 hats for 1068 dollars, gaining thereby 1 dollar on each hat, what did each hat cost him?

(23) If the sum of 103, 29, and 267 be divided by 19, and the quotient be multiplied by 57, and the product be diminished by 197, what will the remainder be?

(24) 8 lambs are worth 16 dollars, and 15 sheep are worth 60 dollars; how many of such sheep ought to be given in exchange for 840 of such lambs?

(25) The sum of the product of two numbers and 355 is eighty-seven thousand four hundred and three; one of the numbers is 216; find the other number.

(26) What number must 416 be multiplied by to produce 154979552?

(27) What number subtracted 28 times from 479632 will leave 20 as a remainder?

(28) A farmer bought 29 bullocks for 1885 dollars, and after keeping them for 3 months, and spending on each 5 dollars per month, he sold all the bullocks for 2610 dollars; what was his gain on each bullock?

24. *If the Divisor terminate with a cypher or cyphers, the process of Division can be shortened by the following Rule.*

RULE. Cut off the cypher or cpyhers from the divisor, and as many figures from the right-hand of the dividend, as there are cyphers to cut off at the right-hand end of the divisor; then proceed with the remaining figures according to the Rule, Art. 21 or Art. 22. as the case may be; and to the last remainder affix the figures cut off from the dividend for the true remainder.

Ex. 1 Divide 57 by 20.

2,0)5,7
2 rem. 1.
10 + 7 or 17.

57 = 50 + 7; now 20 goes 2 in 50 with rem. 10, ∴ when the 5 is divided by the 2, the rem. 1 is really 1 ten, or 10, and the true rem. =

SIMPLE DIVISION.

Ex. 2. Divide 46431 by 500.

5,00)464,31
92 rem. 4.

$46431 = 46400 + 31$, and 46400 divided by $500 = 92$ with rem. 400, ∴ when the 464 is divided by the 5, the rem. 4 is really 400, and the true rem. 431.

Ex. 3. Divide 375340 by 5900.

59,00)3753,40(63
354
———
213
177
———
36

∴ quotient $= 63$, and rem. $= 3640$.

Ex. 4. Divide 563854 by 10, by 1000, and by 100000. We may write down the quotient and remainder for each question at once.

Thus: 1st quotient $= 56385$, and rem. $= 4$.
2nd .... $=$ 563, ... $= 854$.
3rd .... $=$ 5, ... $= 63854$.

Ex. XIX.

(1) Divide 34, 43, 56, 80, 135, 260, 1504, by 10, 20, and 30.

(2) Divide 237, 840, 673, 291, 6019, 7820, 81229, 327800, by 40, 60, 70, 100, and 200.

(3) Divide 79048, 6870, 890061, by 240, 1000, 1500, and 2600; and 830678103490 by 100000000.

(4) 806753245 ÷ 9067.

(5) 612709066 ÷ 70602.

(6) 60005836 ÷ 896.

(7) 70867509 ÷ 9986.

(8) 8673456954 ÷ 868.

(9) 200006783 ÷ 93256.

(10) Multiply 14609 by 719 and divide the product by 8067.

(11) How many regiments of 1000 men, and also of 1200 men, can be formed out of one million one hundred thousand men?

(12) Add together twenty-five millions seven hundred and sixty thousand and thirty-four, 75211379 and 4637862; subtract ten millions and seventy-five from the sum; divide the remainder by 100000.

SECTION II.

MONEY TABLES.

CANADIAN CURRENCY.

25. The silver coins are: a 5 cent piece.
                               a 10 " "
                               a 20 " "
                               a 25 " "
                               a 50 " "

100 cents make one dollar, or $1.

Note 1. The cent which is made of bronze, is one inch in diameter, and 100 cents weigh one pound avoirdupois.

HALIFAX OR OLD CANADIAN CURRENCY.

26.   2 Farthings make 1 Half-penny, or $\frac{1}{2}$d.
        2 Half-pence .... 1 Penny .... 1d.
       12 Pence .... 1 Shilling ...... 1s.
        5 Shillings ...... 1 Dollar ......$1.
        4 Dollars .... 1 Pound ......£1.

Note 2. The farthing is written thus, $\frac{1}{4}$d ; and three farthings thus, $\frac{3}{4}$d.

ENGLISH OR STERLING CURRENCY.

27.   2 Farthings make 1 Half-penny, or $\frac{1}{2}$d.
        2 Half-pence .... 1 Penny .... 1d.
      12 Pence .... 1 Shilling ...... 1s.
      20 Shillings ...... 1 Pound ......£1.

The sovereign, a gold coin = 20 shillings.
The guinea, a gold coin not now in use = 21 shillings.

Note 3. The sterling pound = 4.86\frac{2}{3}$ Canadian currency

UNITED STATES CURRENCY.

28.   10 Mills .... make 1 Cent.
            10 Cents .... 1 Dime.
            10 Dimes .... 1 Dollar.
            10 Dollars ...... 1 Eagle.

WEIGHTS AND MEASURES.

TABLE OF TROY WEIGHT.

29. Troy Weight is used in weighing gold, silver, diamonds, and other articles of a costly nature; and also in determining specific gravities.

24 Grains, gr........make 1 Pennyweight .... 1 dwt.
20 Pennyweights........ 1 Ounce .... 1 oz.
12 Ounce .... 1 Pound........... 1 lb. or 1 ℔.

TABLE OF AVOIRDUPOIS WEIGHT.

30. Avoirdupois Weight is used in weighing all heavy articles, which are coarse and drossy, or subject to waste, as butter, meat, and the like, and all objects of commerce, with the exception of medicines, gold, silver, and some precious stones.

16 Drams, dr.........make 1 Ounce............ 1 oz.
16 Ounces............... 1 Pound............ 1 lb.
25 Pounds............... 1 Quarter .... 1 qr.
4 Quarters, or 100 lbs..... 1 Hundredweight .. 1 cwt.
20 Hundredweights ...... 1 Ton............. 1 ton.

Note. 1 lb. Avoirdupois weighs 7000 grs. Troy.

TABLE OF APOTHECARIES' WEIGHT.

31. Apothecaries' Weight is used in mixing medicines.

20 Grains, gr.........make 1 Scruple .... 1 sc. or 1 ℈
3 Scruples .... 1 Dram ...... 1 dr. or 1 ℨ
8 Drams .... 1 Ounce ...... 1 oz. or 1 ℥
12 Ounces................ 1 Pound ...... 1 lb. or 1 ℔

TABLE OF LINEAL MEASURE.

32. In this measure, which is used to measure distances, lengths, breadths, heights, depths, and the like, of places or things:

12 Lines ....make 1 Inch ...... 1 l.
12 Inches .... 1 Foot ...... 1 ft.
3 Feet, or 36 in. .... 1 Yard ...... 1 yd.
6 Feet .... 1 Fathom .... 1 fth.
5½ Yards, meaning 5 yards and } 1 Rod, Pole, }
  a half yard................ }   or Perch } 1 po.
40 Poles, or 220 yds. .... 1 Furlong.... 1 fur.
8 Furlongs, or 1760 yds. .... 1 Mile ...... 1 mi.
3 Miles .... 1 League .... 1 lea.

The following measurements may be added, as useful in certain cases:

4 Inches make 1 Hand (used in measuring horses).
22 Yards ...... 1 Chain }
100 Links ...... 1 Chain } (used in measuring land.)

A degree is equal to 60 geographical, or nearly 69½ English miles.

TABLE OF CLOTH MEASURE.

33. In this measure, which is used by linen and woollen drapers:

2¼ Inches make 1 Nail .... 1 nl.
4 Nails ...... 1 Quarter .... 1 qr.
4 Quarters ... 1 Yard .... 1 yd.
5 Quarters ... 1 Ell (English).
6 Quarters ... 1 Ell (French).

TABLE OF SQUARE MEASURE.

34. This measure is used to measure all kinds of surface or superficies, such as land, paving, flooring, in fact everything in which length and breadth are to be taken into account.

A SQUARE is a four-sided figure, whose sides are equal, each side being perpendicular to the adjacent sides. See figure below.

A square inch is a square, each of whose sides is an inch in length; a square yard is a square, each of whose sides is a yard in length.

144 Square Inches make 1 Square Foot...1 sq. ft. or 1 ft.
9 Square Feet .... 1 Square Yard ..1 sq. yd. or 1 yd.
30¼ Square Yards ...... 1 Square Pole ..1 sq. po. or 1 po.
40 Square Poles ...... 1 Square Rood ..1 ro.
4 Roods .... 1 Acre .... 1 ac.

25000 Square Links = 1 Rood.
100000 .... = 1 Acre.
10 ...... Chains = 1 Acre.
4840 ...... Yards = 1 Acre.
640 .... Acres = 1 Square Mile.

Note. This table is formed from the table for lineal measure, by multiplying each lineal dimension by itself.

The truth of the above table will appear from the following considerations.

Suppose AB and AC to be lineal yards placed perpendicularly to each other.

Then $ABCD$ is a square yard. If AE, EF, FB, AG, GH, HC, each $= 1$ lineal foot, it appears from the figure that there are 9 squares in the square yard, and that each square is 1 square foot.

The same explanation holds good of the other dimensions.

TABLE OF SOLID OR CUBIC MEASURE.

35. This measure is used to measure all kinds of solids, or figures which consist of three dimensions, length, breadth, and depth or thickness.

A CUBE is a solid figure contained by six equal squares; for instance, a die is a cube. A cubic inch is a cube whose side is a square inch. A cubic yard is a cube whose side is a square yard.

1728 Cubic Inches.......... make 1 Cubic Foot, or 1 c. ft.
  27 Cubic Feet .... 1 Cubic Yard, or 1 c. yd.
  40 Cubic Feet of Rough Timber or
  50 Cubic Feet of Hewn Timber 1 Load.
  42 Cubic Feet .... 1 Ton of Shipping.
 128 Cubic Feet of Fire-wood .... 1 Cord.
  16 Cubic Feet of Fire-wood .... 1 Cord-foot.

The truth of the first part of above table will appear from the following considerations.

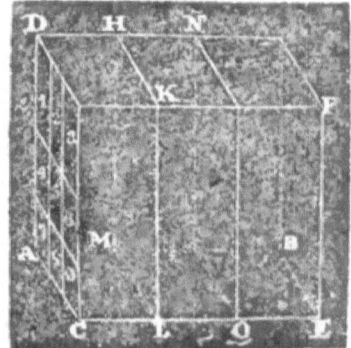

If AB, AC, and AD be perpendicular to each other, and each of them a lineal yard in length, then the figure DE is a cubic yard.

Suppose DH a lineal foot, and $HKLM$ a plane drawn parallel to side DC.

By the table Art. 34, there are 9 square feet in side DC. There will therefore be 9 cubic feet in the solid figure DL.

Similarly, if another lineal

foot *HN* were taken, and a plane *NO* were drawn parallel to *HL*, there would be 9 cubic feet contained in the solid figure *HO*.

Similarly, there would be 9 cubic feet in the solid figure *NE*.

Therefore, there are 27 cubic feet in the solid figure *DE*, or in 1 cubic yard.

Note. A pile of wood 4 feet high, 4 feet wide, and 8 feet long, makes a cord.

MEASURES OF CAPACITY.

TABLE OF WINE MEASURE.

36. In this measure, by which wines and all liquids, with the exception of malt liquors and water, are measured

```
 4 Gills  ......make 1 Pint ......1 pt.
 2 Pints  ..........1 Quart ....1 qt.
 4 Quarts  ........1 Gallon ... 1 gal.
63 Gallons  ........1 Hogshead..1 hhd.
 2 Hogsheads......1 Pipe......1 pipe.
 2 Pipes  ..........1 Tun ......1 tun.
```

TABLE OF ALE AND BEER MEASURE.

37. In this measure, by which all malt liquors and water are measured:

```
 2 Pints..........make 1 Quart ....1 qt.
 4 Quarts ............1 Gallon ....1 gal.
 9 Gallons ............1 Firkin ....1 fir.
18 Gallons ............1 Kilderkin .1 kil.
36 Gallons ............1 Barrel ....1 bar.
1½ Barrels, or 54 Gallons 1 Hogshead .1 hhd.
 2 Hogsheads......... 1 Butt......1 butt.
 2 Butts .............1 Tun...... 1 tun.
```

TABLE OF DRY MEASURE.

38.
```
 2 Pints  ...........make 1 Quart ....1 qt.
 4 Quarts ................1 Gallon ...1 gal.
 2 Gallons ................1 Peck......1 pk.
 4 Pecks..................1 Bushel ....1 bu.
36 Bushels................1 Chaldron..1 ch.
```

39. 34 Pounds make 1 Bushel of Oats.
    48 Pounds .... 1 Bushel of B'kwheat, Barley or Timothy
    50 Pounds .... 1 Bushel of Flax Seed.
    56 Pounds .... 1 Bushel of Rye or Indian Corn.
    60 Pounds .... 1 Bushel of Wheat, Potatoes, Peas, Beans, Onions, or Red Clover Seed.

Note 1. Grains are sold by the cental (100 lbs.), or by parts thereof.

MEASURES OF TIME.

TABLE OF TIME.

40. 1 Second is written thus 1″.
  60 Seconds........make 1 Minute ....1′.
  60 Minutes.... .... 1 Hour.................1 hr.
  24 Hours.............. 1 Day ....1 day.
   7 Days .... 1 Week ....1 wk.
   4 Weeks, or 28 days .. 1 Lunar month ....1 mo.
  365 Days .... 1 Civil or common year.1 yr.

Note 2. 60 minutes make 1 degree, or 60′ make 1°.
    A degree is the 360th part of the circumference of a circle.

A year is divided into 12 months, called Calendar Months, the number of days in each of which may be easily remembered by means of the following lines;

    Thirty days hath September,
    April, June and November:
    February has twenty-eight alone,
    And all the rest have thirty-one:
    But leap-year coming once in four,
    February then has one day more.

Note 3. A civil or common year = 52 wks, 1 day.
    A leap year   = 366 days.

Every year which is divisible by 4 without a remainder is a LEAP OR BISSEXTILE YEAR; except those years which complete a century (i. e. a hundred years), the numbers expressing which century, are *not* divisible by four; thus 1600 and 2000 are leap years, because 16 and 20 are exactly divisible by 4; but 1700, 1800 and 1900 are not leap years, because 17, 18, and 19 are not exactly divisible by 4.

ARITHMETIC.

MISCELLANEOUS TABLE.

41.
12 Units ...... make	1	Dozen.
12 Dozen	1	Gross.
12 Gross............	1	Great Gross.
20 Units...........	1	Score.
24 Sheets of Paper....	1	Quire.
20 Quires	1	Ream.
100 Pounds	1	Quintal.
196 Pounds	1	Barrel of Flour.
200 Pounds	1	Barrel of Pork or Beef.

Note. A sheet folded into two leaves is called a folio, into 4 leaves a quarto, into 8 leaves an octavo, into 16 leaves a 16 mo, into 18 leaves an 18 mo, &c.

REDUCTION.

42. When a number is expressed in one or more denominations, the method of finding its value in one or more other denominations is called REDUCTION. Thus, £1 is of the same value as 240 d., and 7 s. 1½ d. is of the same value as 342 farthings, and conversely; the method or process by which we find this to be so, is REDUCTION.

43. First. *To express a number of a higher denomination or of higher denominations in units of a lower denomination.*

RULE. Multiply the number of the highest denomination in the proposed quantity by the number of units of the next lower denomination contained in one unit of the highest, and to the product add the number of that lower denomination, if there be any in the proposed quantity.

Repeat this process for each succeeding denomination, till the required one is arrived at.

Ex. 1. How many cents in $75.65 **cents?**

By the Rule,

$75.65
100
———
7500 + 65 = 7565 cents.

Reason. Since 100 cents make one dollar; $75 = (75 × 100 cts.) = 7500 cts., ∴ $75.65 = 7500 + 65 = 7565 cents.

∴ $75.65 = 7565 cents.

REDUCTION.

Ex. 2. Reduce £2 to farthings.
By the Rule,

$$\begin{array}{c} £ \\ 2 \\ 20 \\ \hline 40s. \\ 12 \\ \hline 480d. \\ 4 \\ \hline 1920q. \end{array}$$

Reason for the Rule.

$£1 = 20s., \therefore £2 = (2 \times 20)s. = 40s.$

$1s. = 12d., \therefore 40s. = (40 \times 12)d. = 480d.$

$1d. = 4q., \therefore 480d. = (480 \times 4)q. = 1920q.$

$\therefore £2 = 40s. = 480d. = 1920q.$

Ex. XX.

Reduce

(1) £709. 16s., 8d, to farthings.
(2) 17 mls., 1 fur., 2 ft., 6 in. to inches.
(3) 8 tons, 2 cwt., 3 qrs., 5 lbs. to drams.
(4) 612 ac., 2 r., 27½ yds. to square inches.
(5) 10 mls. 5 fur., 5 po., 5 yds., 0 ft., 5 in., 5 ls. to lines
(6) 5 ac., 3 per., 29 yds. to square inches.
(7) 17 days to minutes.
(8) 2 lbs., 11 oz., 20 grs. to grains.
(9) 2 lea., 2 mls., 7 fur. to yards.
(10) 23 cub. yds., 1000 in. to cubic inches.
(11) 13 galls., 3 qts. to gills.
(12) 220 bushels to quarts.
(13) 3 yrs., 315 days to minutes.
(14) 27 lbs., 5 oz., 16 dwts. to grains.
(15) 47 lbs., 11 oz., 6 drs., 2 sc. to grains.
(16) £200. 17s., 8½d. to half-pence.
(17) 219 ac., 2 r., 16 per. to square yards.
(18) 218 yds., 2 qrs., 3 nls. to nails.
(19) £2376. 19s., 8½d. to farthings.
(20) 216 cwt., 2 qrs., 17 lbs. to pounds.
(21) 25° 36′ to seconds.
(22) 8 mls., 3 fur., 4 yds. to inches.
(23) £312 17s., 6½d. to farthings.
(24) 105 lbs. Troy to grains.

ARITHMETIC.

(25) 26 English ells to nails.
(26) 37 French ells to nails.
(27) £567. 0s. 6¼d. to farthings.
(28) 287 lbs., 6 oz. to scruples.
(29) 3 pipes to gallons.
(30) £200. 19s., 6½d. to farthings.

44. *Secondly*. *To express a number of lower denomination or denominations in units of a higher denomination.*

RULE. Divide the given number by the **number** of units which connect that denomination with the next higher, and the remainder, if any, will be the number of surplus units **of the lower denomination.**

Carry on this process, till **you arrive at the denomination** required.

Ex. 1. How many tons, cwts., &c., are there in 27658 drams?
By the Rule,

$$16 \begin{cases} 2 \\ 8 \end{cases} \begin{array}{|l} 27658 \\ \hline 13829 - 10 \text{ drs.} \end{array}$$

16 drs. = 1 oz., ∴ 27658 ÷ 16 = 1728 oz. + 10 drs.

$$16 \begin{cases} 2 \\ 8 \end{cases} \begin{array}{|l} 1728 \\ \hline 864 - 0 \text{ oz.} \end{array}$$

16 oz. = 1 lb., ∴ 1728 oz. ÷ 16 = 108 lbs. + 0 oz.

$$25 \begin{cases} 5 \\ 5 \end{cases} \begin{array}{|l} 108 \\ \hline 21 - 3 \end{array}$$

25 lbs. = 1 qr., ∴ 108 lbs. ÷ 25 = 4 qrs. + 8 lbs.

$$4 \begin{array}{|l} 4 - 1 \\ \hline 1 \end{array}$$ 8 lbs. 4 qrs. = 1 cwt., ∴ 4 qrs ÷ 4 = 1 cwt. + 0 qrs.

cwt., 0 qrs., 8 lbs., 0 oz., 10 drs.

∴ 27658 drams = 1 cwt., 0 qrs., 8 lbs., 0 oz., 10 drs.

Ex. 2. In 17392 cents, how many dollars and cents.
By the Rule,

$$100 \begin{cases} 10 \\ 10 \end{cases} \begin{array}{|l} 17392 \\ \hline 1739 - 2 \\ \$173 - 92 \text{ cts.} \end{array}$$

Reason for the Rule.
100 cents = $1, ∴ 17392 cts. ÷ 100
= $173 + 92 cts., ∴ 17392 cents
= $173.92 cts.

Note. From the above example, we see that by cutting off the last 2 figures on the right of any number of cents, gives the dollars, and the figures so cut off will be the cents.

Ex. XXI.

Reduce.
(1) 123290 farthings to pounds.
(2) 13172 grs. to lbs. Troy.
(3) 18191 pts. to gallons.
(4) How many leagues in 76787568 inches?
(5) How many tons, &c., in 2007008 drams?
(6) How many acres in 93827 perches?
(7) In 167812 grs., how many lbs. Troy?
(8) In 8756765637 lines, how many miles, &c.?
(9) In 7678678956 drs., how many tons, &c.?
(10) In 121605 in., how many miles, &c.?
(11) In 98006 grs., how many lbs. Troy, &c.?
(12) In 2022752 drs., how many tons, &c.?
(13) How many lbs., ozs., drs., &c., in 702917 grs.?
(14) How many years (365 ds.), &c., in 1727893 seconds?
(15) How many acres, &c., in 172425 yards?
(16) How many yards in 13856832 cubic inches?
(17) How many acres in 1244160000 sq. inches?
(18) How many yards, &c., in 500 nails?
(19) In 131075 seconds, how many degrees, &c.?
(20) In 31557600 seconds, how many days, &c.?
(21) In 219612 pts., how many hogsheads of beer?
(22) In 300738 pts., how many hogsheads of wine?
(23) In 912715 lbs., how many bushels of wheat?
(24) In 1000000 lbs. of oats, how many bushels?
(25) In 7263 lbs. of timothy seed, how many bushels?
(26) In 30747 cents, how many dollars?
(27) How many pounds, &c., in 973647 farthings?

COMPOUND ADDITION.

45. Compound Addition is the method of collecting several numbers of the same kind, but containing different denominations of that kind, into one sum.

Rule. Arrange the numbers, so that those of the same denomination may be under each other in the same column, and draw a line below them.

Add the numbers of the lowest denominations together, and find by Reduction how many units of the next higher denomination are contained in this sum.

Write the remainder, if any, under the column just added, and carry the quotient to the next column.

Proceed thus with all the columns.

Ex. 1. Add together $21.97, $28.76, $38.39.

By the Rule,

$21.97
$28.76
$38.39
──────
$89.12

The sum of the right-hand column is 22; write 2 under that column, and carry 2 to the next; the sum of the next column together with the 2 carried is 21; write 1 under that column and carry 2 to the next, and so on; the same way as was done in the Simple Rules, and for the same reason.

Ex. 2. Find the sum of £6. 6s., £3. 13s. 0¾d., £33. 15s. 11½d., and £43. 0s. 8¼d.

£	s.	d.
6	6	0
3	13	0¾
35	15	11½
43	0	8¼
£88	15	8½

$1q. + 2q. + 3q. = 6q. = 1\tfrac{1}{2}d.$ ∴ write down ½d., and carry 1d.

Then $1d. + 8d. + 11d. = 20d. = 1s. 8d.$; write down 8d., and carry 1s.

Then $1s. + 15s. + 13s. + 6s. = 35s. = £1.$ 15s.; write down 15s, and carry £1.

Then £1 + £43 + £35 + £3 + £6 = £88; write down £88.

Note. The method of proof in the Compound Rules is the same as in the Simple Rules.

Ex. XXII.

Add together,

(1)
$26.79
$39.17
$28.63

(2)
£	s.	d.
6	9	8
8	10	4
5	12	3

(3)
qrs.	lbs.	oz.
2	17	12
6	24	13
1	6	8

(4)
lbs.	oz.	dwt.	gr.
35	3	4	12
27	8	14	22
41	9	17	10
2	3	13	21

(5)
lbs.	oz.	dr.	sc.	gr.
17	8	2	1	5
12	10	6	0	19
6	6	4	2	18
17	11	7	2	19

COMPOUND SUBTRACTION.

(6) $ 286.97
    6126.35
     517.68
    9612.07
     712.15

(7)
tons	cwt.	qrs.	lbs.	oz.
21	16	2	24	10
26	5	1	22	9
1	17	3	19	12
19	12	0	18	9
218	10	1	12	8

(8)
yds.	qrs.	nls.
27	2	3
35	3	2
217	1	3
89	2	2
207	3	2

(9)
mls.	fur.	per.	yds.	ft.
2	3	8	2	2
25	7	21	4	1
3	6	23	2	0
17	4	19	3	2
29	5	16	1	1

(10)
£	s.	d.
38	6	7¼
29	16	8½
39	17	6¾
21	18	7
15	17	8

(11)
dys.	hrs.	min.	sec.
2	16	16	17
27	22	22	33
19	21	30	37
28	23	39	50
36	20	45	55

(12) $2219.64
    3812.75
     913.25
     837.19
     687.29

(13)
ac.	ro.	per.	yds.	ft.	in.
5	0	7	13	2	5
7	3	9	22	8	107
9	1	16	29	2	96
19	2	22	27	6	108
0	3	7	28	3	12

(14)
tons	cwt.	qrs.	lbs.	oz.	drs.
23	15	2	20	5	0
21	17	0	24	1	13
43	19	3	24	15	15
3	9	2	17	13	11
6	6	1	0	7	8

(15) $5617.28
     208.09
     516.99
    3712.89
     984.75

COMPOUND SUBTRACTION.

46. COMPOUND SUBTRACTION is the method of finding the difference between **two numbers** of the same kind, but containing different **denominations** of that kind.

RULE. Place **the less number below the** greater, so that **the numbers** of **the same** denomination may be under each other in **the** same column, **and** draw a line below them.

Begin at the right hand, and subtract if possible each number of the lower line from that which stands above it, and set the remainder underneath.

But when any number in the lower line is greater than the number above it, add to the upper one as many units of the same denomination as make one unit of the next higher denomination; subtract as before, and carry one to the number of the next higher denomination in the lower line.

Proceed thus throughout the columns.

Ex. 1. From £51. 0s. 8½d., take £47. 18s. 7¾d.

By the Rule,

```
   £    s.   d.
  51 .  0 .  8½
  47 . 18 .  7¾
  ─────────────
  £3 .  2 .  0¾
```

Method of working. I cannot take 3q. from 2q., so I add 1d., or 4q., to the 2q., making it 6q.; then 3q. from 6q. leaves 3q.; write down the 3q.; in order to increase the lower number equally with the upper, I add 1d. to the 7d., making it 8d.; then 8d. from 8d. leaves 0d.; write down 0d. I work the remaining columns in the same way, and find the required answer.

Ex. 2. From $978.29 take $678.93.

```
  $978.29
  $678.93
  ───────
  $299.36
```

This example is worked in the same way as Simple Subtraction.

Ex. XXIII.

```
        £    s.   d.                      lbs.  oz. drs. sc.  grs.
(1)    33 . 17 .  4              (2)      27 .  8 .  6 . 2 . 15
       18 .  8 . 10                       17 .  9 .  3 . 1 . 19
       ─────────────                      ─────────────────────────

       lbs.  oz. dwt.                     mls. fur. per. yds. ft.
(3)    12 .  6 .  3              (4)      25 .  6 . 32 .  4 .  2
        9 .  7 . 16                       22 .  7 . 37 .  3 .  2
       ─────────────                      ─────────────────────────

       yds. qrs. nls.  in.                c. yds.  c. ft.  c. in.
(5)   106 .  1 .  2 .  1         (6)      325 .   22 .    101
       92 .  3 .  3 .  1¼                 296 .   25 .    386
       ─────────────────                  ─────────────────────────

       ac.  ro.  per.  yds.  ft.  in.     wks. dys. hrs. min. sec
(7)    29 .  2 . 27 .  29 .  2 .  6  (8)   7 .  5 .  6 . 36 . 17
       27 .  3 . 29 .  27 .  8 .  8        6 .  6 . 20 . 46 . 20
       ──────────────────────────────     ─────────────────────────
```

COMPOUND MULTIPLICATION.

(9)
£	s.	d.
129	16	8¼
75	18	9½

(10)
cwt.	qrs.	lbs.	oz.	drs.
7	2	15	6	12
6	3	24	10	14

(11) $2967.78
    1898.89

(12)
cords.	c. ft.
193	107
97	125

(13) $325.68
     297.99

(14)
ac.	ro.	per.	yds.	ft.	in.
297	1	23	2	1	101
189	2	28	2¼	2	127

(15)
c. yds.	c. ft.	c. in.
278	3	1127
198	8	1478

(16)
mls.	fur.	per.	yds.	ft.	in.
117	0	27	5	1	9
89	7	38	4	2	11

(17)
degs.	min.	sec.
29	29	38
22	49	59

(18)
tons.	cwt.	qrs.	lbs.	oz.	drs.
293	16	1	21	6	15
287	19	2	22	11	14

(19)
yds.	qrs.	nls.	in.
1209	1	1	1
1198	2	2	1¼

(20)
bu.	pk.	gal.	qt.
268	2	1	1
197	3	1	3

(21)
bu.	pk.	gal.	qt.
19672	0	1	1
18998	3	1	3

COMPOUND MULTIPLICATION.

47. COMPOUND MULTIPLICATION is the method of finding the amount of any proposed compound number, that is, of any number composed of different denominations, but all of the same kind, when it is repeated a given number of times.

RULE. Place the multiplier under the lowest denomination of the multiplicand.

Multiply the number of the lowest denomination by the multiplier, and find the number of units of the next denomination contained in this first product; if there be a remainder, write it down; for the second product, multiply the number of the next denomination in the multiplicand by the multiplier, and after adding to it the above-mentioned number of units, proceed with the result as with the first product.

Carry this operation through with all the different denominations of the multiplicand.

Multiplier not greater than 12.

ARITHMETIC.

Ex. 1. Multiply £1. 14s. 9¾d. by 11.

$$\begin{array}{ccc} £ & s. & d. \\ 1 & .\ 14 & .\ 9¾ \\ & & 11 \\ \hline £19 & .\ 2 & .\ 11¼ \end{array}$$

3¾ × 11 = 33¾. = 8¼d.; write down ¼d.; then 9d. × 11 + 8d. = 99d. + 8d. = 107d. = 8s. 11d.; write down 11d.; then 14s. × 11 + 8s. = 154s. + 8s. = 162s. = £8. 2s.; write down 2s.; then £1 × 11 + £8 = £19; write down £19.

Ex. 2. Multiply $27.78 by 9.

$$\begin{array}{r} \$27.78 \\ 9 \\ \hline \$250.02 \end{array}$$

In this example we do the same as in Simple Multiplication, observing to place the point separating the dollars and cents in its proper place.

Ex. XXIV.

(1) £   s.   d.
    12 . 9 . 6
              2

(2) lbs. oz. drs. sc.
     17 . 5 . 6 . 2
                  3

(3) lbs. oz. dwt. grs.
     18 . 6 . 5 . 10
                   4

(4) yds. qrs. nls.
     27 . 3 . 3
              5

(5) mls. fur. per. yds. ft.
     27 . 7 . 26 . 4 . 2
                       6

(6) $237.19
           7

(7) cwt. qrs. lbs. oz. drs.
     16 . 0 . 17 . 0 . 15
                        8

(8) mls. fur. per. yds. ft. in.
     6 . 4 . 6 . 2 . 1 . 9
                         9

(9) $609.93
          10

(10) wks. dys. hrs. min.
     7 . 5 . 18 . 16
                  11

(11) ac. ro. per. yds. ft. in.
     7 . 3 . 29 . 20 . 1 . 108
                             12

(12) £   s.   d.
     20 . 17 . 7¾
               7

(13) lbs. oz. drs. sc.
     74 . 11 . 5 . 2
                  12

(14) bu. pk. qt.
     7 . 3 . 1
             3

(15) dys. hrs. min.
     2 . 3 . 59
            10

(16) gal. qt. pt.
     4554 . 3 . 1
               11

(17) dys. hrs. min. sec.
     365 . 5 . 48 . 57
                    12

(18) £   s.   d.
     73 . 17 . 8¾
               11

(19) ac. ro. per.
     14 . 3 . 39
             9

(20) 297.68
         12

COMPOUND MULTIPLICATION.   59

```
        bu.  pk. gal.              £   s.   d.                 lbs.  oz.  dwt.  grs.
(21)  2782 . 2 . 1      (22)  70 . 0 . 11½       (23)  18 .  3 . 14 .  5
                9                            12                                   12
```

```
                                      £    s.    d.
(24)  $917.75           (25)  17 . 15 . 0¾       (26)  $1875.25
           8                             9                          12
```

```
                              mls.  fur.  per.  yds.
             (27)  54 .  3 .  18 .  5
                                  7
```

If the Multiplier be a composite number, each of whose factors is less than 12, multiply by one of them, and the resulting product by another, and so on. The last product so obtained, is the required product.

Find the product of 2 cwt., 3 qr., 17 lbs. by 63.

```
  cwt. qrs.  lbs.
    2 .  3 . 17        The factors of 63 are 9 and 7. First, we
               9       multiply by 9 and the product we get by 7;
   26 .  1 .  3        which clearly is the same as multiplying 2
               7       cwt., 3 qr., 17 lbs. by 63.
  183 .  3 . 21           *Note.* The same result is obtained, by
                       taking the factor 7 first, and then the 9.
```

Ex. XXV.

```
       ac.  ro.  per           mls.  fur.  per.            £    s.   d.
(1)   56 .  2 .  9      (2)   27 .  6 .  9       (3)  19 . 11 .  4
            28                           54                              14¼
```

```
       lbs.  dwt.  grs.         £    s.    d.            yds.  qr.  nls.  in.
(4)   21 . 13 . 17      (5)  17 . 11 . 8¼        (6)  27 .  1 .  3 .  2
              77                         20                                54
```

```
      cwt  qrs.  lbs.  oz.  drs.        £    s.   d.
(7)   2 .  3 . 23 . 12 .  6    (8)  72 . 19 . 9½    (9)  $209.18
                      63                         81                    35
```

```
       c. yds.  c. ft.  c. in.       lbs.  oz.  dwt.  grs.          £    s.    d.
(10)   17 .  21 .  57    (11)  3 .  8 . 15 . 13    (12)  42 . 10 . 9¼
                 84                            49                             88
```

	dys. hrs. min. sec.		lbs. oz. drs. sc.
(13)	5 . 17 . 39 . 20	(14)	74 . 11 . 5 . 2
	120		84

	lbs. oz. dwt. grs		£ s. d.
(15)	6 . 2 . 3 . 17	(16)	13 . 7 . 4¾
	5382		275

	ac. ro. per. yds. ft. in.		mls. fur. per. yds. ft. in.
(17)	20 . 2 . 17 . 15 . 3 . 3	(18)	2 . 6 . 2 . 3 . 0 . 5
	64		375

	£ s. d.				bu. pk. gal.
(19)	2 . 6 . 8¼	(20)	$237.15	(21)	10 . 2 . 1
	900		500		800

When the Multiplier is not a Composite number and larger than 12, the easiest method will be to split the number into factors and parts:

Thus, $29 = 4 \times 7 + 1$; $19 = 6 \times 3 + 1$; $39 = 12 \times 3 + 3$.

Ex. 1. Multiply £2579. 0s. 0¾d. by 2331.
$2331 = 2000 + 300 + 30 + 1$.
   $= 1000 \times 2 + 100 \times 3 + 10 \times 3 + 1$.
   $= 10 \times 10 \times 10 \times 2 + 10 \times 10 \times 3 + 10 \times 3 + 1$.

```
          £         s.    d.
       3579  .     0  .  0¾    for 1
                  10
       ─────────────────
      25790  .     0  .  7½    for 10
                  10
     ─────────────────
     257900  .     6  .  3     for 10 × 10, or 100
                  10
    ─────────────────
    2579003  .     2  .  6     for 100 × 10, or 1000
                   2
    ─────────────────
    5158006  .     5  .  0     for 1000 × 2, or 2000.
add  773700  .    18  .  9     for £257900. 6s. 3d. × 3, or for 300.
add   77370  .     1  . 10½    for £25790. 0s. 7½d. × 3, or for 30.
add    2579  .     0  .  0¾    for 1.
    ─────────────────
    6011656  .     5  .  8¼    for 2000 + 300 + 30 + 1, or 2331.
```

Ex. XXVI.

(1) cwt. qrs. lbs. oz.
    3 . 3 . 21 . 5
    ──────────────
         89

(2) lbs. oz. dwt. grs.
    6 . 2 . 3 . 17
    ──────────────
        5463

(3) £ s. d.
    2 . 6 . 9½
    ──────────
        938

(4) cwt. qrs. lbs. oz. drs.
    2 . 3 . 23 . 6 . 7
    ──────────────────
           627

(5) £   s. d.
    4 . 18 . 9½
    ───────────
        561

(6) lbs. oz. drs. sc. grs.
    15 . 2 . 3 . 2 . 7
    ──────────────────
           712

(7) If a man gets $2.25 a day, how much will that be in 209 days?

(8) When wheat is selling for $1.27 a bushel, how many dollars will a farmer get for a load of 52 bushels of wheat?

(9) A butcher had an ox weighing 1625 lbs., live weight, at 6 cents a pound how much will he have to pay altogether?

(10) A boiler-builder bought 29 boiler plates, each weighing 1 qr., 17 lbs., 8 oz., what was the weight of the whole of them?

(11) If the Government of Ontario sells one hundred thousand acres of wild land for forty cents an acre, how many dollars will it obtain for the whole?

COMPOUND DIVISION.

48. COMPOUND DIVISION is the method of dividing a compound number, that is, a number composed of several denominations, but all of the same kind, into as many equal parts as the divisor contains units; and also of finding how often one compound number is contained in another of the same kind.

When the Divisor is a number either larger, or not larger than 12.

RULE. Place the numbers as in Simple Division; then find how often the divisor is contained in the highest denomination of the dividend; put this number down in the quotient; multiply as in Simple Division and subtract.

If there be a remainder, reduce that remainder to the next

inferior denomination, adding to it the number of that denomination in the dividend, and repeat the division.

Carry on this process through the whole dividend.

When the Divisor is less than 12.

Ex. 1. Divide £676. 19s. 9½d. by 11.

```
      £     s.    d.
11 | 676 . 19 .  9½
     ─────────────────
      61 . 10 . 10½
```

£676 ÷ 11 gives £61 as a quotient and £5 over; £5 + 19s. = 119s., 119s.
rem. 8q. ÷ 11 gives 10s. as a quotient and 9s. over; 9s. + 9d. = 117d., 117d. ÷ 11 gives 10d. as a quotient and 7d. over; 7d. + 2q. = 30q., 30q. ÷ 11 gives 2 as a quotient and rem. 8q.

When the Divisor is greater than 12 and not a Composite number, the work may stand thus:

Ex. 2. Divide £297. 4s. 8d. by 73.

By the Rule,
```
       £.    s.   d.
73 ) 297 .  4 .  8 ( £4
     292
     ─────
       5
      20  [add the 4s.]
     ─────
73 ) 104 ( 1s.
      73
     ─────
      31
      12  [add the 8d.]
     ─────
73 ) 380 ( 5d.
     365
     ─────
      15
```

We first subtract £4 taken 73 times, i.e. £292 from £297. 4s. 8d., there remains £5. 4s. 8d.

Now £5. 4s. 8d. = 104s. 8d., from this we subtract 1s. taken 73 times, i.e. 73s. from 104s., there remains 31s., ∴ there is 1s. in quotient.

31s. 8d. = 380d., from this we subtract 5d. taken 73 times, i.e. 365d., there remains 15d. over

∴ £4. 1s. 5d. goes 73 times in £297. 4s. 8d., and 15d. over.

∴ the Quotient is £4. 1s. 5d. and 15d. over.

When the Divisor is a Composite number greater than 12, we may divide as in Ex. 1, successively by each factor, and the last quotient so obtained will be the required quotient.

Ex. 3. Divide 975 mls., 3 fur., 24 per. by 56.

Since 56 = 8 × 7, the work may stand thus:

```
       mls.  fur.  per.
   8 | 975 .  3 .  24
   7 | 121 .  7 .  18
       ─────────────────
        17 .  3 .  14
```

Note. The same result would be obtained by dividing first by 7 and then by 8.

COMPOUND DIVISION.

Ex. XXVII.

(1) £278. 15s. 8d. ÷ 5.
(2) 237 lbs., 5 oz., 6 dwt. ÷ 8.
(3) 217 mls., 5 fur., 16 per., 2 yds. ÷ 9.
(4) 115 yds., 2 qrs., 2 nls. ÷ 5.
(5) 865 lbs., 9 oz., 2 sc., 10 grs. ÷ 6.
(6) £2078. 17s. 1¼d. ÷ 11.
(7) 67 tons, 13 cwt., 1 qr., 17 lbs ÷ 27.
(8) 976 ac., 2 ro., 19 per., 25 yds. ÷ 56.
(9) 612 cwt., 17 lbs., 2 drs. ÷ 705.
(10) 8627 mls., 6 fur., 2 yds. ÷ 1247.
(11) 612 bu., 2 pks., 1 gal. 2 qts. ÷ 96.
(12) £2851. 16s. 4½d. ÷ 54.
(13) 247 lbs., 10 oz., 7 drs., 1 sc. ÷ 57.
(14) 200 mls., 3 fur., 6 per. ÷ 211.
(15) 416 ac., 3 ro., 19 per., 7 yds. ÷ 318.
(16) 614 tons, 2 cwt., 3 qrs. ÷ 564.
(17) 917 c. yds., 9 c. ft., 100 c. in. ÷ 163.
(18) 926 lbs., 5 oz., 3 drs., 2 sc. ÷ 212.
(19) 3068 lbs., 8 dwt. ÷ 634.
(20) £1914. 10s. 5d. ÷ 758.
(21) £215. 12s. 6¼d. ÷ 317.
(22) 125 yrs., 127 dys., 16 hrs., 47 min. ÷ 397.
(23) $2267.84 ÷ 267.
(24) $5693.75 ÷ 425.

(25) If a person earned $600 a year, how much is that a day? How much per day, omitting the Sundays?

Note. A year = 365 days.

(26) A farm of 57 acres is let for $265.05, for a year; how much is that for an acre?

(27) A farmer sold 57 bushels of wheat for $65.55; how much did he get for one bushel?

(28) The annual rent of a house is $132; how much must be put aside every week so as to have the whole rent ready at the end of the year?

ARITHMETIC.

When the divisor and dividend are both compound numbers of the same kind.

RULE. Reduce both numbers to the same denomination. Divide as in Simple Division. The Quotient will be the answer required. Ex. 1. How often is 3s. 7d. contained in £8. 15s. 7d.?

```
  3s. 7d.      £8. 15s. 7d.
   12             20
   ──            ───
   43            175
                  12
                  ──
                 2107
             43)2107(49
                 172
                 ───
                 387
                 387
```

Reason for the Rule.

3s. 7d. = 43d., £8. 15s. 7d. = 2107d.; 43d. subtracted 49 times from 2107d. leaves no remainder.

∴ 49 times is the answer.

Ex. 2. I employ twice as many men as women, the wages of the former are 3s. 6d. each, and of the latter 1s. 10d. each per day. The weekly wages amount to £23. 17s. How many men, and how many women do I employ?

£23. 17s. ÷ 6 = £3. 19s. 6d. = 954d. = am^t. of daily wages
Daily wages of 2 men and 1 woman = 3s. 6d. × 2 + 1s. 10d.
$\qquad\qquad\qquad\qquad$ = 8s. 10d. = 106d.

```
106)954(9
    954
    ───
```
∴ there are 18 men and 9 women.

Ex. XXVIII.

Divide,

(1) £684. 7s. 6d. by £76. 0s. 10d.
(2) £171. 1s. 10½d. by £57. 0s. 7½d.
(3) 9 lbs., 9 oz., 3 dwt. 12 grs. by 5 dwt. 9 grs.
(4) 4 mls., 1 fur., 2 yds. by 1 ml., 3 fur., 2 ft.
(5) 6 cwt. 2 qrs. by 1 qr., 3 oz.
(6) 12 lbs., 6 oz., 2 sc., by 1 lb., 6 oz., 2 sc., 10 grs.
(7) 3 yds., 1 qr., 2 nls. by 1 qr., 2 nls.
(8) 1 dy., 1 hr., 12 min. by 1 hr. 3 min.
(9) 5 sq. per., 7 yds., 108 in. by 2 yds. 1 ft.
(10) $141.05 by $2.17.
(11) $221 by $2.21.

49. *To reduce old Canadian to the Decimal Currency.*

RULE. Multiply the pounds by 4, the product is dollars. Multiply the shillings by 20, the product is cents.
Reduce the pence to farthings and add the given farthings, if any; then multiply by 5 and divide by 12, the quotient is cents.
The sum of these results is the answer required.
How many dollars and cents in £72. 19s. 9½d. ?

$$£1 = \$4, \therefore £72 = \$72 \times 4 = \$288.00$$
$$1s. = 20 \text{ cts.}, \therefore 19s. = 19 \times 20 \text{ cts} = 3.80$$
$$9\tfrac{1}{2}d. = 38q., \therefore 38q. \times 5 \div 12 = 190 \div 12 = 15\tfrac{10}{12}$$
$$\overline{\$291.95\tfrac{10}{12}}$$

Therefore the required answer is $291.95\tfrac{10}{12}$.

Ex. XXIX.

How many dollars and cents in

(1) £25. 6s. 3d. (2) £57. 19s. 3d.
(3) £207. 17s. 8d. (4) £153. 18s. 5d.
(5) £217. 17s. 0d. (6) £319. 15s. 7½d.
(7) £612. 19s. 11¼d. (8) £63. 9s. 9¾d.
(9) £912. 12s. 6d. (10) £711. 5s. 5¼d.
(11) £1117. 0s. 7½d. (12) £47. 7s. 9d.
(13) £2017. 6s. 8d. (14) £75. 9s. 8½d.
(15) £37. 18s. 7½d. (16) £87. 13s. 9d.

50. *To reduce dollars and cents to Halifax or old Canadian Currency.*

RULE. Divide the dollars by 4, the quotient is pounds.
If there is any remainder bring it to cents and add the given cents if any; then divide by 20, the quotient is shillings.
If any cents are left, multiply them by 3 and divide by 5; the quotient is pence. By arranging these several quotients properly, the required answer is obtained.
How many pounds, shillings and pence in $1279.12½ ?

4 | 1279.12½ $3 + 12½ cts. = 300 cts. + 12½ cts. =
———————   312½ cts. ; 312½ cts. ÷ 20 = 15s. and
£319 and $3 over.   12½ cts. over; 12½ cts. × 3 ÷ 5 = 7½d.

Therefore the answer is £319. 15s. 7½d. The above is evidently correct; because $4 = £1, 20 cts. = 12d., 5 cts. = 3d.

Ex. XXX.

How many **pounds**, shillings and pence in

(1)	$217.25	(2)	$327.55	(3)	$17.33
(4)	$84.50	(5)	$75.95	(6)	$125.87½
(7)	$867.87½	(8)	$1162.40	(9)	$1393.62½
(10)	$1937.20	(11)	$2220.29	(12)	$3785.48

Ex. XXXI.

MISCELLANEOUS EXAMPLES.

PAPER I.

(1) The population of the counties on the river St. Lawrence in 1861 was as follows: Leeds, thirty-five thousand seven hundred; Grenville, twenty-four thousand one hundred and ninety-one; Dundas, eighteen thousand seven hundred and seventy-seven; Stormont, eighteen thousand one hundred and twenty-nine; Glengarry, twenty-one thousand one hundred and eighty-seven. Find the total population of these five counties.

(2) By the census of 1848, the population of Montreal was fifty-five thousand one hundred and forty-six; of Toronto, twenty-three thousand five hundred and three; of Hamilton, nine thousand eight hundred and eighty-nine; of Ottawa, six thousand two hundred and seventy-five; of Kingston, eight thousand three hundred and sixty-nine; of London, four thousand five hundred and eighty-four. Find the whole population of those cities.

(3) Add, one hundred thousand, two hundred and twenty-nine thousand seven hundred and thirteen, fifty-eight thousand seven hundred and five, six hundred and twelve thousand five hundred and seventeen, nine hundred and ninety-nine thousand nine hundred and ninety-nine, eight hundred and thirty-three thousand seven hundred and nineteen, seven hundred and sixty-eight thousand three hundred and nine, fifty thousand and fifty.

(4) Add, five thousand and five, seven thousand and eighteen, seventeen thousand nine hundred and fifteen, twenty-eight thousand seven hundred and nineteen, nine thousand and twelve, eight hundred and seven thousand five hundred and twelve, seven hundred and seventeen thousand and seventeen, ninety-three thousand five hundred and two, two hundred and twelve thousand six hundred and seven.

MISCELLANEOUS EXAMPLES.   67

(5) How many miles in 178006 inches?

(6) In 1848 the value of the imports into Canada was $8375180.20; in 1861, the value of the imports was $43054836; the population at the former date was 1493332, at the latter 2506755. Find 1st., the value of the imports for each person in 1848 and in 1861, and 2nd., the difference between these values.

PAPER II.

(1) What is the price of 818 bushels of wheat at 8s. $10\frac{1}{2}d.$ per bushel?

(2) A farmer sold 67 bushels of wheat at $1.62 a bushel; bought a suit of clothes for $18, 82 yards cotton at $13\frac{1}{2}$ cents a yard, a stove for $16. How much was left of the price of the wheat?

(3) If a Government was to divide 72812 acres equally among 397 discharged soldiers, how much would each receive?

(4) A farmer brought 160 bushels of wheat to mill when wheat was worth $1.60 per bushel, and in exchange got 27 barrels of flour. How much was he charged for the flour per barrel?

(5) A merchant has a piece of cloth containing $42\frac{1}{2}$ yards, worth 6s. $6\frac{1}{2}d.$ a yard. How many dresses of $8\frac{1}{2}$ yards each can be made out of it, and what will each cost?

(6) A farmer sold in the Toronto market 618 barrels of flour for £1. 13s. 9d. per barrel; and bought 84 yards of cotton at 17 cents a yard, 5 lbs. tea at 3s. 9d. a lb., 2 tons of coal at £1. 15s. per ton, 8 sheep at £2. 11s. 9d. each, 15 head of cattle at £12. 19s. 9d. each. How much can he deposit in a bank allowing that he takes $50 home with him?

PAPER III.

(1) In one year there were coined in the British mint 203761 pounds of gold, value £9520732. 14s. 6d. Required the value of each pound?

(2) Three persons bought a ship for $63000; the first taking one share, the second three, and the third five. How much do they severally pay?

(2) If a contribution of £354. 11s. 6d. is made up in equal shares by 26 men, how much must each give?

(4) What is the 29th part of 10 ac., 2 ro., 7 per., 2 yds?

E

(5) Divide 300 tons, 15 cwt., 3 qrs., equally among 347 men. How much will each get?

(6) Soldiers marching in quick time, make 110 steps in a minute, each step 2 ft. 6 in. long. In what time would a company of soldiers march 20 miles in quick time, allowing half an hour for rest?

PAPER IV.

(1) Add together £6. 17s. 6d., $30.27, £3. 12s. 9d. $75.83; giving your answer in decimal currency.

(2) Three boys went out together to fish, the first caught eight, the second as many and three more, the third as many as his two comrades all but one. How many did each of the last two boys catch?

(3) Three boys, Thomas, William, and Alexander, had between them 6 cents; Thomas had one, William two, and Alexander three; they bought fifty-four marbles with their money. How many ought each boy to get?

(4) Four men went out one night to fish, borrowing both boat and nets. A man was to have 4 shares of the catch as often as the owner of the net was to have one; but, a man was to have only two shares as often as the owner of the boat had one. The catch was four barrels of herrings. What was each party's share in dozens; each barrel containing 38 dozens of herrings?

(5) It is found by observation that in each square inch of the human skin there are about 1000 pores; and the surface of the body of a middle sized man contains about 2304 square inches, or 16 square feet. Required, the number of pores in the surface of such a body, 999 being supposed to be contained in each square inch?

(6) The sum of two numbers is 84889; the difference between them is 889. What are the numbers?

PAPER V.

(1) Find the product of 72678397 and 86073?

(2) The quotient is 73697; the remainder 3687; the divisor 11689. Find the dividend?

(3) The minuend is twenty-seven thousand eight hundred and twelve; the difference, fifteen thousand nine hundred and eight. Find the subtrahend?

(4) There are seven addends all equal; their sum is eighty-nine thousand two hundred and sixty-four. Find one of them?

(5) In the census of 1861, Rutland contained twenty-two thousand nine hundred and eighty-three inhabitants; Northamptonshire, ninety-six thousand eight hundred and one; Huntingdonshire, sixty-four thousand one hundred and eighty-three; Leicestershire, ninety-one thousand three hundred and eight; Nottinghamshire, one hundred and ninety thousand and sixty. What was the sum of the population of the above five counties in 1861?

(6) During the Crimean war, out of the French army there were killed in action or missing ten thousand two hundred and forty; drowned in a wreck, seven hundred and four; died of various diseases before the battle of Alma, eight thousand and eighty-four; died of disease before Sebastopol, four thousand three hundred and twelve; died in hospitals, &c., seventy-two thousand two hundred and forty-seven. How many were lost altogether?

PAPER VI.

(1) In 1861 the population of Edinburgh was 160302; of Glasgow, 168795 more than that of Edinburgh; of Aberdeen 71973; of Inverness, 24527 more than that of Aberdeen. What was the total population of all these places in 1861?

(2) The paid up capital of each of the following Banks doing business in Ontario, is: of the Bank of Montreal, $6000000; Bank of British North America, $4866666; of Quebec Bank, $1467750; of Bank of Toronto, $800000; of Ontario Bank, $1909640; of Royal Canadian Bank, $590382; of Merchants' Bank, $862033. Find the total amount of the paid up capital of the above named Banks?

(3) The amount of revenue, from the named sources during 1866, was as follows: Customs, $7328146.68; Excise, $1888576.76; Postage, $621936.42; Public-works, $417474; Education, $66554; Common School Fund, $122142.77. Find the whole revenue from these sources?

(4) A person has $975. He buys a team for $375, a wagon for $82, a plough for $16, a stove $16, a reaping machine for $153, 12 sheep for $8 each, 2 cows $25 each, 3 pigs $6 a piece, pays his servantman 3 months' wages at $20 a month, and the rest he lays out in flour at $1.75 per 100 pounds. How many pounds of flour will he have?

(5) Among 635 men divide equally 86895 acres.

(6) How many inches in 10 mls., 3 per., 4 yds.?

SECTION III.

GREATEST COMMON MEASURE.

51. A MEASURE of any given number is a number which will divide the given number exactly, *i. e.* without a remainder.

Thus, 2 is a measure of 6, because 2 is contained 3 times exactly in 6.

52. A MULTIPLE of any given number is a number which contains it an exact number of times. Thus, 6 is a multiple of 2.

53. A COMMON MEASURE of two or more given numbers is a number which will divide each of the given numbers exactly. Thus, 3 is a common measure of 18, 27, and 36.

The GREATEST COMMON MEASURE (G. C. M.) of two or more given numbers, is the greatest number which will divide each of the given numbers exactly. Thus, 9 is the greatest common measure of 18, 27, and 36.

54. *To find the greatest common measure of two numbers.*

RULE. Divide the greater number by the less.

If there be a remainder, divide the first divisor by it.

If there be still a remainder, divide the second divisor by this remainder, and so on; always dividing the last preceding divisor by the last remainder, till nothing remains.

The last divisor will be the greatest common measure required.

Ex. Find the G. C. M. of 144 and 240.

By the Rule,

```
144 ) 240 ( 1
      144
      ‾‾‾
      96 ) 144 ( 1   bringing down last divisor 144 for a dividend.
           96
           ‾‾
           48 ) 96 ( 2  ..  ..  ..  ..  ..  96  ..  ..  ..
                96                              
                ‾‾           ∴ 48 is G. C. M. required.
```

LEAST COMMON MULTIPLE.

Ex. XXXII.

Find the G. C. M. of

(1) 8 and 18.   (2) 6 and 15.   (3) 4 and 22.
(4) 16 and 28.   (5) 20 and 32.   (6) 24 and 39.
(7) 26 and 44.   (8) 30 and 42.   (9) 36 and 56.
(10) 46 and 116.   (11) 58 and 174.   (12) 315 and 378.
(13) 366 and 128.   (14) 180 and 210.   (15) 310 and 630.
(16) 1216 and 424.   (17) 127 and 445.   (18) 6408 and 7264.
(19) 3042 and 3094.   (20) 7040 and 7392.
(21) 1441 and 1572.   (22) 46436 and 23025.
(23) 21168 and 204624.   (24) 97482 and 29579.
(25) 828597 and 738140.   (26) 326337 and 737800.

LEAST COMMON MULTIPLE.

55. A COMMON MULTIPLE of two or more given numbers is a number which will contain each of the given numbers an exact number of times. Thus, 144 is a common multiple of 3, 9, 18, and 24.

The LEAST COMMON MULTIPLE (L. C. M.) of two or more given numbers is the least number which will contain each of the given numbers an exact number of times. Thus, 72 is the least common multiple of 3, 9, 18, and 24.

56. *When the least common multiple of several numbers is required, the most convenient practical method is that given by the following Rule.*

RULE. Arrange the numbers in a line from left to right, with a comma placed between every two.

Divide those numbers which have a common measure by that common measure, and place the quotients so obtained and the undivided numbers in a line beneath, separated as before.

Proceed in the same way with the second line, and so on with those which follow, until a row of numbers is obtained in which there are no two numbers which have any common measure greater than unity.

Then the continued product of all the divisors and the numbers in the last line will be the least common multiple required.

Note. It will in general be found advantageous to begin

with the lowest prime number 2 as a divisor, and to repeat this as often as can be done; and then to proceed with the prime numbers 3, 5, &c., in the same way.

Ex. 1. Find the L. C. M. of 10, 12, and 16.
By the Rule,

```
2 | 10, 12, 16
2 |  5,  6,  8
        5,  3,  4
```

$10 = 2 \times 5$, $12 = 2 \times 2 \times 3$, $16 = 2 \times 2 \times 2 \times 2$.

∴ L. C. M. must clearly contain as factors
2×5 for 10.
$2 \times 5 \times 2 \times 3$ for 10 and 12.
$2 \times 5 \times 2 \times 3 \times 2 \times 2$ for 10, 12, and 16.

∴ L. C. M. $= 2 \times 2 \times 5 \times 3 \times 4 = 240$.

Note. The process of finding the L. C. M. may often be shortened by striking out in the same line every number which exactly measures any other number in that line.

Ex. 2. Find the L. C. M. of 9, 14, 16, 18, 24, 36, and 38.

```
2 | 9, 14, 16, 18, 24, 36, 38
2 |    7,  8,     12, 18, 19
2 |    7,  4,      6,  9, 19
       7,  2,      3,  9, 19
```

Every multiple of 36 must be a multiple of 9 and of 18; ∴ strike out 9 and 18: for the same reason strike out 3 in the 4th line.

∴ L. C. M. $= 2 \times 2 \times 2 \times 7 \times 2 \times 9 \times 19 = 19152$.

Ex. XXXIII.

Find the L. C. M. of

(1) 2, 4, and 10.
(2) 8, 9, and 12.
(3) 12, 16, and 18.
(4) 20, 28, and 36.
(5) 16, 24, and 30.
(6) 24, 56, and 84.
(7) 15, 25, and 105.
(8) 6, 33, 24, and 32.
(9) 7, 21, 6, 14, and 25.
(10) 7, 8, 9, 10, and 12.
(11) 24, 28, 36, 22, and 16.
(12) 2, 5, 45, 15 and 25.
(13) 9, 4, 8, 15, and 27.
(14) 15, 20, 24, 21, and 3
(15) 4, 5, 7, 8, 15, 21, and 30.
(16) 2, 7, 9, 13, 15, 52, and 63.
(17) 3, 7, 21, 11, 77, and 198.
(18) 100, 56, 35, 125, and 150
(19) 22, 55, 19, 15, 95, and 133.
(20) 48, 64, 27, 33, 110 and 165.

SECTION IV.

FRACTIONS.

57. Let unity be represented by the line AB, which we will consider to be one yard in length.

Suppose AB to be divided into 3 equal parts AD, DE, EB; then one of such parts AD is a foot or one-third part of the yard, and it is denoted thus $\frac{1}{3}$ (read *one-third*); two of them AE, or two feet, thus $\frac{2}{3}$ (read *two-thirds*); three of them AB, or three feet, or the whole yard, thus $\frac{3}{3}$ or 1.

If another equal portion BF of a second yard BC, divided in the same manner as the first, be added, then AF, or four feet, is denoted thus $\frac{4}{3}$; and so on.

Such expressions, representing any number of the equal parts of a unit, *i. e.* of a quantity which is denoted by 1, are called BROKEN NUMBERS OR FRACTIONS.

58. A FRACTION denotes one or more of the equal parts of a unit; it is expressed by two numbers placed one above the other with a line between them; the lower number is called the DENOMINATOR (Den^r.), and shews into how many equal parts the unit is divided; the upper is called the NUMERATOR (Num^r.), and shews how many of such parts are taken to form the fraction.

59. A Fraction also represents the quotient of the num^r. by the den^r.

Thus $\frac{2}{3} = 2 \div 3$; for we obtain the same result, whether we divide one unit, AB or 1 yard, into three equal parts AD, DE, EB, each $= 1$ ft. or 12 in., and take two of such parts AE (represented by $\frac{2}{3}$), $= 12$ in. × $2 = 24$ in., or divide 2 units, AC or 2 yards, into three equal parts, AE, EF, FC, each $= 2$ ft. or 24 in., and take one of such parts AE; which is equal to $\frac{1}{3}$rd part of AC or 2 units, or $= 2 \div 3$. Hence $\frac{2}{3}$ and $2 \div 3$ have the same meaning.

60. When fractions are denoted in the manner above explained, they are called VULGAR FRACTIONS.

ARITHMETIC.

61. Fractions, whose den^r. are composed of 10, or **of 10** multiplied by itself any number of times, are called DECIMAL FRACTIONS, or DECIMALS.

VULGAR FRACTIONS.

62. In treating of the subject of Vulgar Fractions, **it is** usual to make the following distinctions:

(1) A PROPER FRACTION is one whose num^r. is less than the den^r.; thus $\frac{3}{4}$, $\frac{1}{5}$, $\frac{2}{7}$, are proper fractions.

(2) AN IMPROPER FRACTION is one whose num^r. is equal to or greater than the den^r.: thus, $\frac{5}{5}$, $\frac{8}{3}$, $\frac{7}{4}$ are improper fractions.

(3) A SIMPLE FRACTION is one whose num^r. and den^r. are simple integer numbers; thus, $\frac{1}{3}$, $\frac{3}{4}$ are simple fractions.

(4) A MIXED NUMBER is composed of a whole number and a fraction; thus $5\frac{1}{3}$, $7\frac{2}{3}$ are mixed numbers, representing respectively 5 units, together with $\frac{1}{3}$th of a unit; and 7 units, together with $\frac{2}{3}$ths of a unit.

(5) A COMPOUND FRACTION is a fraction of a fraction; thus $\frac{1}{2}$ of $\frac{3}{4}$, $\frac{2}{3}$ of $\frac{7}{8}$ of $\frac{9}{10}$, are compound fractions.

(6) A COMPLEX FRACTION is one which is either a fraction of a mixed number in one or both terms of the fraction; thus, $\dfrac{\frac{3}{4}}{\frac{2}{7}}$, $\dfrac{2\frac{1}{2}}{3}$, $\dfrac{3}{4\frac{2}{3}}$, $\dfrac{2\frac{1}{4}}{5\frac{1}{2}}$, $\dfrac{\frac{3}{4} \text{ of } \frac{1}{2}}{2\frac{1}{3}}$ are complex fractions.

63. It is clear from what has been said, that every whole number or integer may be considered as a fraction whose den^r. is 1; thus, $5 = \frac{5}{1}$, for the unit is divided into 1 part comprising the whole unit, and 5 of such parts, that is 5 units, are taken.

64. *To multiply a fraction by a **whole** number.*

RULE. Multiply the numerator by the whole number.

$$\frac{2}{5} \times 2 = \frac{2 \times 2}{5} = \frac{4}{5}$$

For in $\frac{2}{5}$ and $\frac{4}{5}$, the unit is divided into 5 equal parts, **and** *twice* as many parts are taken in $\frac{4}{5}$ as are taken in $\frac{2}{5}$.

Ex. XXXIV.

Multiply (1) $\frac{2}{3}$ and $\frac{11}{12}$ each **separ**ately by 2, 3, 5, 7, 9, and 12; and (2) $\frac{23}{31}$ and $\frac{98}{107}$ each separately by 6, 8, 11, 106 and 157.

VULGAR FRACTIONS.

65. *To divide a fraction by a whole number.*

RULE. Multiply the denominator by the whole number.

$\frac{2}{5} \div 2 = \frac{2}{5 \times 2} = \frac{2}{10}$. The value of each part in $\frac{2}{5}$ is *twice* as large as the value of each part in $\frac{2}{10}$; but the same number of parts are taken in each, $\therefore$ $\frac{2}{5}$ is *twice* as large as $\frac{2}{10}$, or $\frac{2}{5} \div 2 = \frac{2}{10}$.

Ex. XXXV.

Divide (1) $\frac{2}{3}$ and $\frac{4}{7}$ each separately by 2, 3, 5, 6, 9, and 12; and (2) $\frac{13}{9}$ and $\frac{27}{11}$ each separately by 3, 5, 11, 56, and 100.

66. If the numerator and denominator of a fraction be *both* multiplied, or *both* divided, by the same number, the value of the fraction will not be altered.

$\frac{3}{4} = \frac{3 \times 2}{4 \times 2} = \frac{6}{8}$ Since $8 = 4 \times 2$, *two* of the parts in $\frac{6}{8}$ are equivalent to *one* of the parts in $\frac{3}{4}$; but since $6 = 3 \times 2$, there are *twice* as many parts taken in $\frac{6}{8}$ as there are in $\frac{3}{4}$, therefore $\frac{3}{4} = \frac{6}{8}$. In figure, Art. 57, AE represents either $\frac{1}{3}$rd or $\frac{2}{6}$ths of AC.

67. Hence it follows that a whole number may be converted into a vulgar fraction with any required den'., by multiplying the number by the required den'. for the num'. of the fraction, and placing the required den'. underneath.

For $5 = \frac{5}{1}$, and to convert it into a fraction with a den'., 6 or 17, we have $5 = \frac{5}{1} = \frac{5 \times 6}{1 \times 6} = \frac{30}{6}$; also $5 = \frac{5}{1} = \frac{5 \times 17}{1 \times 17} = \frac{85}{17}$.

Ex. XXXVI.

Reduce (1) 3, 5, 8, 15, to fractions with denⁿˢ. 2, 9, and 13; and (2) 9, 12, 17, 37, to fractions with denⁿˢ. 8, 10, and 57.

68. *To represent an improper fraction as a whole or mixed number.*

RULE. Divide the numerator by the denominator.

If there be no remainder, **the quotient** will be a whole number.

If there **be a** remainder, put down the quotient as the integral part, **and the** remainder **as** the num'. of the fractional part, and the given den'. as the den'. of the fractional part.

Ex. Reduce $\frac{24}{4}$ and $\frac{24}{5}$ to whole or mixed numbers.

By the Rule,

$$\frac{24}{4} = 6. \quad \text{For } \frac{24}{4} = \frac{4 \times 6}{4 \times 1} = \frac{6}{1} \text{ (Art. 66)} = 6.$$

$$\frac{24}{5} = 4\tfrac{4}{5}. \quad \text{For } \frac{24}{5} = \frac{20 + 4}{5} = \frac{20}{5} + \frac{4}{5} = 4 + \frac{4}{5} = 4\tfrac{4}{5}.$$

Ex. XXXVII.

Express the following improper fractions as mixed or whole numbers:

(1) $\tfrac{6}{2}$.   (2) $\tfrac{8}{3}$.   (3) $\tfrac{1.9}{4}$.   (4) $\tfrac{1.9}{5}$.   (5) $\tfrac{1.9}{6}$.
(6) $\tfrac{1.9}{7}$.   (7) $\tfrac{2.0}{8}$.   (8) $\tfrac{1.9}{9}$.   (9) $\tfrac{9.1}{10}$.   (10) $\tfrac{1.1.2}{14}$.
(11) $\tfrac{2.9.1}{27}$.   (12) $\tfrac{9.0.4}{47}$.   (13) $\tfrac{1.0.0.9}{107}$.   (14) $\tfrac{2.0.6.4.0}{260}$.   (15) $\tfrac{1.7.1.3}{133}$.

69. *To reduce a mixed number to an improper fraction.*

RULE. Multiply the whole number or integer by the denominator of the fraction, and to the product add the numerator of the fractional part.

The result will be the required num^r., and the den^r. of the fractional part the required den^r.

Ex. Convert $3\tfrac{3}{4}$ into an improper fraction.
By the Rule,

$$3\tfrac{3}{4} = \frac{3 \times 4 + 3}{4} = \frac{12 + 3}{4} = \frac{15}{4}.$$

$$\text{For } 3\tfrac{3}{4} = \frac{3}{1} + \frac{3}{4} = \frac{3 \times 4}{1 \times 4} + \frac{3}{4} = \frac{12}{4} + \frac{3}{4} = \frac{12 + 3}{4} = \frac{15}{4}.$$

Ex. XXXVIII.

Reduce the following mixed numbers to improper fractions:

(1) $1\tfrac{1}{3}$.   (2) $2\tfrac{1}{17}$.   (3) $1\tfrac{1}{15}$.   (4) $17\tfrac{3}{8}$.   (5) $12\tfrac{6}{7}$.
(6) $203\tfrac{4}{5}$.   (7) $2\tfrac{11}{14}$.   (8) $29\tfrac{7}{8}$.   (9) $704\tfrac{12}{13}$.
(10) $900\tfrac{31}{101}$.   (11) $5\tfrac{7}{510}$.   (12) $53\tfrac{7}{33}$.   (13) $21\tfrac{3}{1750}$.
(14) $148\tfrac{237}{465}$.   (15) $13\tfrac{1938}{2133}$.   (16) $25\tfrac{289}{7100}$.   (17) $197\tfrac{205}{3034}$.

70. *To reduce a compound fraction to its equivalent simple fraction.*

RULE. Multiply the several numerators together for the numerator of the simple fraction, and the several denominators together for its denominator.

VULGAR FRACTIONS.

Ex. 1. Convert $\frac{2}{3}$ of $\frac{5}{6}$ into a simple fraction.

By the Rule,
$$\frac{2}{3} \text{ of } \frac{5}{6} = \frac{2 \times 5}{3 \times 6} = \frac{10}{18}.$$

For $\frac{2}{3}$ of $\frac{5}{6}$ = twice $\frac{1}{3}$ of $\frac{5}{6}$ = twice $\frac{5}{6} \div 3$ = twice $\frac{5}{18}$ (Art. 65)

$$= \frac{5 \times 2}{18} \text{ (Art. 64)} = \frac{10}{18}.$$

Note 1. Before applying the above Rule, mixed numbers must be reduced to improper fractions.

Note 2. In reducing compound fractions to simple ones, we may strike out from *any* num^r. and *any* den^r. such factors as are common to both; for this is in fact simply dividing the num^r. and den^r. of a fraction by the same number. (Art. 66.)

Ex. 2. Reduce $\frac{3}{5}$ of $2\frac{1}{12}$ of $1\frac{1}{15}$ to a simple fraction.

$\frac{3}{5}$ of $2\frac{1}{12}$ of $1\frac{1}{15} = \frac{3}{5}$ of $\frac{25}{12}$ of $\frac{16}{15} = \frac{3 \times (5 \times 5) \times (4 \times 4)}{5 \times (3 \times 4) \times (3 \times 5)}$

$= \frac{\cancel{3} \times \cancel{5} \times \cancel{5} \times \cancel{4} \times 4}{\cancel{5} \times \cancel{3} \times \cancel{4} \times 3 \times \cancel{5}} = \frac{4}{3}$, dividing num^r. and den^r. by 3, 5, 5, 4, factors common to both.

Ex. XXXIX.

Reduce the following compound fractions to simple ones:

(1) $\frac{3}{4}$ of $\frac{4}{5}$. (2) $\frac{1}{2}$ of $1\frac{9}{12}$. (3) $\frac{7}{8}$ of $2\frac{1}{4}$.

(4) $\frac{5}{10}$ of $\frac{3}{11}$. (5) $\frac{5}{8}$ of $2\frac{2}{3}$. (6) $\frac{3}{8}$ of $1\frac{1}{4}$.

(7) $18\frac{2}{3}$ of $5\frac{7}{15}$ of 10. (8) $11\frac{2}{3}$ of $8\frac{1}{7}$.

(9) $\frac{5}{6}$ of $2\frac{1}{3}$ of 9. (10) $\frac{4}{7}$ of $3\frac{2}{7}$ of $3\frac{1}{2}$.

(11) $\frac{5}{6}$ of $\frac{3}{11}$ of $\frac{2}{3}$ of $1\frac{2}{3}$. (12) $\frac{3}{14}$ of $4\frac{2}{3}$ of $\frac{9}{57}$ of $6\frac{8}{11}$ of $\frac{7}{32}$.

(13) $\frac{5}{11}$ of $2\frac{1}{2}$ of $\frac{4}{7}$ of $10\frac{1}{2}$. (14) $\frac{7}{5}$ of $12\frac{1}{2}$ of $\frac{1}{4}$ of $\frac{5}{8}$ of $\frac{3}{8}$ of 9.

(15) $\frac{1}{15}$ of $\frac{7}{5}$ of $\frac{12}{15}$ of $\frac{9}{7}$ of $\frac{3}{15}$ of 2 of $\frac{5}{7}$.

(16) $\frac{2}{3}$ of $\frac{3}{4}$ of $\frac{9}{7}$ of $70\frac{2}{3}$ of $\frac{3}{15}$ of $1\frac{7}{11}$ of 147.

71. A fraction is in its LOWEST TERMS, when its numerator and denominator are PRIME to each other.

72. *To reduce a fraction to its lowest terms.*

Rule. Divide the numerator and denominator by their greatest common measure.

Ex. Reduce $\frac{176}{484}$ to its **lowest** terms.

By the Rule, find the G. C. M. of 176 and 484.

```
176 ) 484 ( 2
      352
      ───
      132 ) 176 ( 1
            132
            ───
            44 ) 132 ( 3
                 132
                 ───
44 ) 176 ( 4    44 ) 484 ( 11
     176             44
     ───             ──
                     44
                     44
                     ──
```

For $\dfrac{176}{484} = \dfrac{44 \times 4}{44 \times 11} = \dfrac{4}{11}$ (Art. 66.)

∴ fraction in its lowest terms $= \dfrac{4}{11}$

Ex. XL.

Reduce each of the following fractions to its lowest terms:

(1) $\frac{2}{4}$. (2) $\frac{10}{15}$. (3) $\frac{14}{21}$. (4) $\frac{12}{15}$.
(5) $\frac{24}{63}$. (6) $\frac{81}{54}$. (7) $\frac{77}{121}$. (8) $\frac{44}{272}$.
(9) $\frac{1428}{7056}$. (10) $\frac{1408}{1081}$. (11) $\frac{875}{1000}$. (12) $\frac{1008}{7056}$.
(13) $\frac{637}{2736}$. (14) $\frac{2008}{8008}$. (15) $\frac{804}{2027}$. (16) $\frac{8104}{10004}$.
(17) $\frac{28111}{35555}$. (18) $\frac{30509}{77140}$. (19) $\frac{4112}{8753}$. (20) $\frac{1284}{1800}$.
(21) $\frac{280711}{350555}$. (22) $\frac{10312}{10312}$. (23) $\frac{7812}{9312}$. (24) $\frac{11286}{10322}$.

73. *To reduce fractions to equivalent ones with a common denominator.*

Rule. Find the **least common** multiple of the denominators; this will be **the** common denominator.

Then **divide the common multiple** so found **by** the denominator of each fraction, and multiply each quotient so found into the numerator of the fraction which belongs to it for the new numerator of that **fraction.**

Note. If the given fractions be in their *lowest* terms, the above rules will reduce them to others having the *least* common denʳ.: if the *least* **common** denʳ. be required, the given fractions should **be** reduced **to their** lowest **terms** before the **rule** is applied.

Ex. 1. Reduce $\frac{1}{2}$, $\frac{1}{4}$, and $\frac{2}{3}$ to equivalent fractions with a common denominator.

VULGAR FRACTIONS.

By the Rule, $12 \mid \dfrac{12, 24, 36}{2,\ 3}$ ∴ L. C. M. $= 12 \times 2 \times 3 = 72$.

∴ the fractions become $= \dfrac{11 \times 6}{12 \times 6} = \dfrac{66}{72}$ (since $72 \div 12 = 6$),

and $\dfrac{17 \times 3}{24 \times 3} = \dfrac{51}{72}$ (since $72 \div 24 = 3$),

and $\dfrac{31 \times 2}{36 \times 2} = \dfrac{62}{72}$ (since $72 \div 36 = 2$),

∴ the required fractions are $\frac{66}{72}$, $\frac{51}{72}$, and $\frac{62}{72}$.

Note. If the den^{rs}. have no common measure, the work will be more quickly done, by multiplying each num^r. into all the den^{rs}., except its own, for a new num^r. for each fraction, and all the den^{rs}. together for the common den^r.

Ex. 2. Reduce $\frac{2}{3}$, $\frac{3}{5}$, and $\frac{5}{7}$ to equivalent fractions with a common den^r.

L. C. M. of the den^{rs}. $= 3 \times 5 \times 7 = 105$.

∴ fract^{ns}. $= \dfrac{2 \times 5 \times 7}{3 \times 5 \times 7},\ \dfrac{3 \times 3 \times 7}{5 \times 3 \times 7},\ \dfrac{5 \times 3 \times 5}{7 \times 3 \times 5}$; or $\dfrac{70}{105},\ \dfrac{63}{105},\ \dfrac{75}{105}$.

Ex. XLI.

Reduce the fractions in each of the following sets to equivalent fractions, having the least common den^r.:

(1) $\frac{3}{4}$ and $\frac{5}{6}$. (2) $\frac{2}{3}$ and $\frac{3}{5}$. (3) $\frac{4}{7}$ and $\frac{7}{8}$.
(4) $\frac{4}{9}$ and $\frac{5}{6}$. (5) $1\frac{1}{8}$ and $3\frac{1}{4}$. (6) $1\frac{1}{2}$ and $2\frac{7}{5}$.
(7) $\frac{7}{10}$ and $1\frac{13}{20}$. (8) $4\frac{4}{5}$ and $6\frac{21}{25}$. (9) $\frac{3}{5}$, $1\frac{1}{2}$, and $\frac{9}{10}$.
(10) $\frac{3}{10}$, $\frac{2}{11}$, and $\frac{5}{6}$. (11) $\frac{7}{15}$, $\frac{1}{2}\frac{1}{4}$, and $\frac{23}{30}$.
(12) $\frac{7}{6}$, $\frac{8}{11}$, $1\frac{2}{3}$, and $\frac{9}{17}$. (13) $1\frac{2}{3}$, $3\frac{3}{4}$, $1\frac{1}{6}$, and $\frac{49}{8}$.
(14) $\frac{7}{12}$, $\frac{7}{10}$, $1\frac{7}{8}$, $1\frac{5}{20}$, and $\frac{7}{15}$. (15) $\frac{1}{2}\frac{3}{8}$, $3\frac{7}{11}$, $\frac{7}{12}$, and $1\frac{5}{8}$.
(16) $\frac{9}{14}$, $1\frac{5}{6}$, $\frac{7}{15}$, $1\frac{3}{5}$, $2\frac{3}{8}$, and $\frac{5}{9}\frac{3}{8}$.
(17) $\frac{3}{8}$, $\frac{5}{6}$, $\frac{9}{4}$, and $1\frac{1}{5}$. (18) $\frac{3}{4}$, $\frac{5}{6}$, $\frac{7}{8}$, and $\frac{9}{10}$.
(19) $\frac{3}{8}$, $\frac{5}{6}$, $\frac{7}{8}$, $\frac{8}{9}$, and $\frac{3}{14}$. (20) $\frac{3}{8}$, $\frac{5}{6}$, $\frac{1}{2}$, and $\frac{3}{11}$.

74. Whenever a *comparison* has to be made between fractions, *in respect of their magnitudes*, they **must** be reduced to equivalent ones with a common den^r.; because then we shall have the unit divided, in the case of each fraction so obtained, into the **same number of equal parts**; and the respective num^{rs}. will shew us how many of such parts are

taken in each case, or which is the greatest fraction, which the next, and so on.

Ex. Which is the greatest, and which the least of the fractions $\dfrac{11 \times 4}{5 \times 9}$, $\dfrac{12 \times 3}{4 \times 10}$, $\dfrac{10 \times 5}{6 \times 8}$, $\dfrac{11 + 4}{5 + 9}$,

The fract$^{ns.}$ in their lowest terms are $\dfrac{44}{45}$, $\dfrac{9}{10}$, $\dfrac{25}{24}$, and $\dfrac{15}{14}$.

L. C. M. of the denr$^{s.}$ = 2520.

∴ the fractions become $\dfrac{44 \times 56}{45 \times 56}$ or $\dfrac{2464}{2520}$, $\dfrac{9 \times 252}{10 \times 252}$ or $\dfrac{2268}{2520}$,

$\dfrac{25 \times 105}{24 \times 105}$ or $\dfrac{2625}{2520}$, $\dfrac{15 \times 180}{14 \times 180}$ or $\dfrac{2700}{2520}$.

∴ $\dfrac{11 + 4}{5 + 9}$ is the greatest, and $\dfrac{12 \times 3}{4 \times 10}$ the least.

Ex. XLII.

Compare the values of

(1) $\tfrac{3}{4}$ and $\tfrac{1}{3}$.   (2) $\tfrac{7}{9}$ and $\tfrac{8}{17}$.   (3) $\tfrac{11}{24}$ and $\tfrac{14}{17}$.

(4) $\tfrac{5}{6}$, $\tfrac{3}{8}$, and $\tfrac{11}{14}$.   (5) $1\tfrac{5}{7}$, $2\tfrac{3}{5}$, and $4\tfrac{8}{9}$.   (6) $\tfrac{5}{8}$, $\tfrac{6}{8}$, and $\tfrac{11}{54}$.

(7) $\tfrac{3}{4}$ of $\tfrac{5}{6}$, $\tfrac{1}{2}$ of $\tfrac{3}{7}$, and $7\tfrac{3}{8}$.   (8) $1\tfrac{1}{5}$, $\tfrac{17}{20}$, $\tfrac{21}{25}$, and $\tfrac{23}{88}$.

(9) $\tfrac{5}{11}$ of $1\tfrac{3}{8}$ of $7\tfrac{1}{4}$, $4\tfrac{1}{2}$ of $\tfrac{3}{13}$, $\tfrac{3}{8}$ of $7\tfrac{1}{2}$ of 11, and $\tfrac{3}{8}$ of $4\tfrac{1}{3}$ of $\tfrac{1}{6}$ of $14\tfrac{7}{11}$.

(10) $\tfrac{911}{783}$ of $\tfrac{85}{171}$, $\tfrac{19}{17}$ of $6\tfrac{1}{2}$ or $\tfrac{11}{56}$ of $1\tfrac{7}{27}$ and $1\tfrac{5}{8}$ of $1\tfrac{3}{8}$ of $5\tfrac{5}{9}$ of $\tfrac{1}{2}$ of $1\tfrac{2}{78}$.

Which is the greater.

(11) $\tfrac{5}{7}$ of a yd. or $\tfrac{3}{8}$ of a yd.

(12) $\tfrac{1}{2}$ of a yd. or $\tfrac{2}{8}$ of a yd.

(13) $1\tfrac{3}{4}$ of $\tfrac{5}{11}$ of $1\tfrac{2}{3}$ of $\tfrac{22}{25}$ of a loaf, or $\tfrac{5}{8}$ of $\tfrac{1}{110}$ of $5\tfrac{1}{2}$ loaves?

ADDITION OF VULGAR FRACTIONS.

75. Rule. reduce the fractions to equivalent ones with the least common denominator.

Add all the new numerators together, and under their sum write the common denominator.

Ex. 1. Find the sum of $\tfrac{1}{2}$, $\tfrac{1}{3}$, and $\tfrac{5}{8}$.

By the Rule,

VULGAR FRACTIONS.

The L. C. M. of the den^{rs}. is 24.

∴ fract^{ns}. become $\dfrac{1 \times 12}{2 \times 12}$ or $\dfrac{12}{24}$, $\dfrac{1 \times 8}{3 \times 8}$ or $\dfrac{8}{24}$, $\dfrac{5 \times 3}{8 \times 3}$ or $\dfrac{15}{24}$.

∴ Their sum $= \dfrac{12 + 8 + 15}{24} = \dfrac{35}{24} = 1\frac{11}{24}$.

Reason for the Rule. In each of the equivalent fractions unity is divided into 24 equal parts, and 12, 8, and 15, of such parts are taken, therefore their sum must be $12 + 8 + 15$, or 35 of such parts, and will be represented by the fraction $\frac{35}{24}$, or by $1\frac{11}{24}$.

Note 1. If the sum of the fractions be a fraction which is not in its lowest terms, reduce it to its lowest terms; and if the result be an improper fraction, then reduce it to a whole or mixed number: thus $\dfrac{147}{105} = \dfrac{49}{35} = 1\frac{14}{35}$: the same remark applies to all results in Vulgar Fractions.

Note 2. Before applying the Rule, reduce all fractions to their lowest terms, improper fractions to whole or mixed numbers, and compound fractions to simple ones.

Note 3. If any of the given numbers be whole or mixed numbers; the whole numbers may be added together as in simple addition, and the fractional parts by the Rule given above.

Ex. 2. Find the sum of $3\frac{5}{12}$, $3\frac{1}{6}$, $2\frac{7}{16}$, and $\frac{3}{4}$ of $3\frac{2}{3}$.

$\frac{3}{4}$ of $3\frac{2}{3} = \frac{3}{4}$ of $\frac{11}{3} = \frac{11}{4} = 2\frac{3}{4}$;

∴ sum of fractions $= 3 + 3 + 2 + 2 + \frac{5}{12} + \frac{1}{6} + \frac{7}{16} + \frac{3}{4}$.

$= 10 + \dfrac{5 \times 4}{12 \times 4} + \dfrac{1 \times 8}{6 \times 8} + \dfrac{7 \times 3}{16 \times 3} + \dfrac{3 \times 12}{4 \times 12}$ (since L. C. M of den^{n}. $= 48$)

$= 10 + \dfrac{20 + 8 + 21 + 36}{48} = 10 + \dfrac{85}{48} = 10 + 1\frac{37}{48} = 11\frac{37}{48}$.

76. The sign () or { }, called BRACKET, enclosing numbers within it, and the sign ― called a VINCULUM, placed over two or more numbers, denotes that all the numbers within the bracket or under the vinculum are equally affected by anything outside the bracket or vinculum, thus $(2 + 3)$ apples or $\overline{2 + 3}$ apples would mean 2 apples + 3 apples, or 5 apples; whereas $2 + 3$ apples would mean 2 units + 3 apples.

Again $\frac{1}{2}+\frac{1}{3}$ of $(2+\frac{1}{4})=\frac{1}{2}+\frac{1}{3}$ of $\frac{9}{4}=\frac{1}{2}+\frac{9}{8}=\frac{4}{8}+\frac{9}{8}=\frac{13}{8}=1\frac{5}{8}$.

$(\frac{1}{2}+\frac{1}{3})$ of $(2+\frac{1}{4}) = (\frac{3}{6}+\frac{2}{6})$ of $(\frac{8}{4}+\frac{1}{4}) = (\frac{5}{6}$ of $\frac{9}{4}=\frac{45}{24}=2\frac{9}{24}$.

$(\frac{1}{2}+\frac{1}{3})$ of $2+\frac{1}{2}=(\frac{3}{6}+\frac{2}{6})$ of $2+\frac{1}{2}=\frac{5}{6}$ of $2+\frac{1}{2}=\frac{10}{6}+\frac{3}{6}=\frac{13}{6}=2\frac{1}{6}$.

Ex. 3. Find the value of $\frac{1}{6}+\frac{1}{3}$ of $(2+\frac{1}{3})+\frac{1}{6}$ of $2\frac{1}{2}+\frac{1}{4}$ of $(\frac{6}{8}+\frac{1}{2})$

value $=\frac{1}{6}+\frac{1}{3}$ of $\frac{7}{3}+\frac{1}{6}$ of $\frac{5}{2}+\frac{1}{4}$ of $(\frac{6}{8}+\frac{4}{8})=\frac{1}{6}+\frac{7}{9}+\frac{5}{12}+\frac{5}{24}$.

$= \dfrac{11}{9} + \dfrac{5}{12} + \dfrac{1}{3} = \dfrac{44+15+12}{36} = \dfrac{71}{36} = 1\frac{35}{36}$.

Ex. XLIII.

Find the sum of,
(1) $\frac{1}{3}$ and $\frac{2}{7}$.   (2) $\frac{3}{4}$ and $\frac{2}{3}$.   (3) 3 and $\frac{1}{3}$.
(4) $\frac{3}{4}$ and $\frac{4}{5}$.   (5) $\frac{5}{12}$ and $\frac{7}{15}$.   (6) $\frac{3}{4}$ and $\frac{7}{14}$.
(7) $\frac{5}{8}$ and $\frac{5}{11}$.   (8) $\frac{3}{8}$ and $\frac{5}{14}$.   (9) $\frac{1}{11}$ and $\frac{5}{13}$.
(10) $1\frac{1}{2}$ and $1\frac{1}{8}$.   (11) $7\frac{2}{8}$ and 8.   (12) $1\frac{1}{3}$ of $2\frac{1}{2}$ and $6\frac{1}{4}$.
(13) $\frac{3}{4}, \frac{4}{5},$ and $\frac{7}{17}$.   (14) $2\frac{3}{8}, \frac{5}{13},$ and $3\frac{1}{12}$.
(15) $6\frac{5}{14}, \frac{1}{3}$ of $1\frac{7}{8},$ and $2\frac{7}{8}$.   (16) $9\frac{1}{2}$ of $2\frac{1}{3}, 1\frac{3}{8},$ and $\frac{4}{7}$.
(17) $\frac{3}{8}, \frac{5}{6},$ and $\frac{1}{8}$ of $(1+1\frac{1}{3})$.

Find the value of,
(18) $\frac{1}{2}+\frac{2}{3}+\frac{3}{4}+\frac{4}{5}$.   (19) $2\frac{1}{2}+3\frac{1}{3}+4\frac{1}{4}+5\frac{1}{5}$.
(20) $5\frac{7}{11}+13\frac{5}{12}+\frac{4}{9}+2\frac{33}{68}$.   (21) $4\frac{5}{6}+\frac{7}{15}+16\frac{9}{10}+25\frac{1}{18}$.
(22) $3\frac{3}{5}+16\frac{7}{3}+7\frac{5}{12}+\frac{2}{3}$ of $3\frac{3}{4}$.
(23) $(2\frac{3}{8}+3\frac{3}{4})$ of $2\frac{5}{11}+3\frac{1}{4}$ of $(16\frac{3}{8}+3\frac{1}{4})+1\frac{3}{8}$ of 11 of $2\frac{1}{12}$.

(24) A gentleman gave £$2\frac{1}{8}$ to A, £$\frac{1}{7}$ to B, £$3\frac{1}{12}$ to C, £$4\frac{1}{8}$ to D, and £$\frac{2}{4}\frac{9}{3}$ to E. How much did he give away?

(25) A man ate $\frac{5}{16}$ of a 4 lb. loaf on Mon., $\frac{1}{12}$ of a similar loaf on Tues., $\frac{1}{16}$ on Wed., $\frac{1}{20}$ on Thurs., $\frac{1}{17}$ on Frid., and on Sat. and Sun. as much as on Mon., Tues., and Wed. How many lbs. of bread did he eat during the week?

SUBTRACTION.

77. Rule. Reduce the fractions to equivalent ones having the least common denominator.

Take the difference of the new numerators, and place the common denominator underneath.

Ex. 1. Subtract $\frac{1}{2}$ from $\frac{5}{8}$.

By the Rule,

The fract$^{ns.}$ become $\dfrac{1 \times 4}{2 \times 4}$ or $\dfrac{4}{8}$, and $\dfrac{5}{8}$,

$\therefore$ their difference $= \dfrac{5-4}{8} = \dfrac{1}{8}$.

Reason for the Rule. In each of the equivalent fractions, unity is divided into 8 equal parts, and there are 5 and 4 parts respectively taken, ∴ the difference must be $5-4$, or $\frac{1}{8}$ of such parts, which is represented by $\frac{1}{8}$.

Note 1. Before applying the Rule, reduce fractions to their lowest terms, improper fractions to whole or mixed numbers, and compound fractions to simple ones.

Note 2. If either of the given fractions be a whole or mixed number, it is most convenient to take separately the difference of the integral parts and that of the fractional parts, and then add the two results together, as in the following examples.

Ex. 2. From $4\frac{3}{8}$ take $2\frac{1}{4}$, or from $(4 + \frac{3}{8})$ take $(2 + \frac{1}{4})$.
Diff^{e}.$= (4 + \frac{3}{8}) - (2 + \frac{1}{4}) = 4 + \frac{3}{8} - 2 - \frac{1}{4}$ (Art. 76.)
$= (4-2) + (\frac{3}{8} - \frac{1}{4}) = 2 + (\frac{3}{8} - \frac{2}{8}) = 2 + \frac{1}{8} = 2\frac{1}{8}$.

Ex. 3. Find the difference between $2\frac{3}{8}$ and $4\frac{1}{4}$.
$\frac{3}{8}$ is greater than $\frac{1}{4}$, and ∴ cannot be taken from it,
∴ we write $4\frac{1}{4}$ thus $(3 + 1 + \frac{1}{4})$, or $(3 + \frac{5}{4})$
then diff^{e}. $= (3\frac{5}{4}) - (2 + \frac{3}{8}) = (3-2) + (\frac{5}{4} - \frac{3}{8}) = 1 + \frac{10}{8} - \frac{3}{8}$
$= 1 + \frac{7}{8} = 1\frac{7}{8}$.

Ex. XLIV.

Find the diff^{s}. between

(1) $\frac{1}{4}$ and $\frac{1}{8}$. (2) $\frac{1}{2}$ and $\frac{1}{3}$. (3) $\frac{3}{4}$ and $\frac{5}{12}$.
(4) $1\frac{2}{5}$ and $\frac{1}{2}\frac{9}{10}$. (5) $3\frac{5}{8}$ and $2\frac{1}{6}$. (6) 7 and $2\frac{9}{16}$.
(7) $10\frac{1}{12}$ and $8\frac{1}{6}$. (8) $17\frac{3}{4}$ and $13\frac{5}{8}$. (9) $1\frac{1}{16}$ and $\frac{3}{4}$.
(10) $4\frac{2}{7}$ and $2\frac{11}{14}$. (11) $15\frac{2}{7}$ and $7\frac{3}{4}$. (12) $20\frac{5}{13}$ and $8\frac{3}{39}$.

(13) A boy ate $\frac{3}{5}$ of a cake, how much less did he leave than he ate?

(14) What number added (1) to $\frac{7}{88}$ will make $1\frac{3}{8}$? and (2) to $2\frac{3}{5}$ will make $8\frac{1}{2}$?

(15) I copied down by mistake $\frac{5}{8}d.$ instead of $\frac{3}{4}d.$, what amount of error did I make?

78. *Examples involving both Addition and Subtraction of Vulgar Fractions.*

Ex. 1. Find the value of $5\frac{1}{4} - 2\frac{1}{2} + \frac{1}{8} + 2\frac{1}{4} - \frac{1}{16}$.
Value $= (5 - 2 + 2) + (\frac{1}{4} - \frac{1}{2} + \frac{1}{8} + \frac{1}{4} - \frac{1}{16})$.
$= 5 + \dfrac{4 - 8 + 2 + 4 - 1}{16} = 5 + \frac{1}{16} = 5\frac{1}{16}$.

ARITHMETIC.

Ex. 2. Find the value of $\frac{2}{3} + \frac{1}{3}$ of $(2-\frac{1}{3}) - \frac{1}{3}$ of $2\frac{1}{2} + \frac{1}{4} - \frac{1}{4}$ of $(\frac{5}{8}-\frac{1}{2})$.

$$\text{Value} = \frac{2}{3} + \frac{1}{3} \text{ of } \left(\frac{6-1}{3}\right) - \frac{1}{3} \text{ of } \frac{5}{2} + \frac{1}{2} - \frac{1}{4} \text{ of } (\frac{5}{8}-\frac{1}{2})$$

$$= \frac{2}{3} + \frac{1}{3} \text{ of } \frac{5}{3} - \frac{1}{3} \text{ of } \frac{5}{2} + \frac{1}{2} - \frac{1}{4} \text{ of } \frac{2}{8} = \frac{2}{3} + \frac{5}{9} - \frac{5}{12} + \frac{1}{2} - \frac{1}{12}$$

$$= 1 - \frac{5}{12} + \frac{1}{2} - \frac{1}{12} = 1 + \frac{1}{2} - \frac{5}{12} - \frac{1}{12} = 1 + \frac{6}{12} - \frac{5}{12} - \frac{1}{12}$$

$$= 1 + \frac{6-5-1}{12} = 1 + \frac{6-6}{12} = 1.$$

Ex. XLV.

Find the value of

(1) $\frac{1}{8} + 2\frac{1}{7} + 13\frac{5}{13} - 3\frac{3}{70}$. (2) $\frac{1}{4} - \frac{3}{8} + \frac{5}{6} - \frac{19}{8}$.

(3) $12\frac{1}{7} - \frac{3}{54} + 7\frac{19}{23} - \frac{1}{3}$ of $\frac{19}{5} + \frac{3}{5}$ of $3\frac{3}{4}$.

(4) $(16\frac{3}{5} - 3\frac{1}{4})$ of $3\frac{1}{2} - 16\frac{3}{5} + 3\frac{1}{4}$ of $3\frac{1}{8}$.

(5) $6\frac{1}{4} + \frac{7}{12}$ of $\frac{9}{10}$ of $3\frac{1}{3} - \frac{45}{76} - 5\frac{3}{4}$.

(6) $6\frac{1}{4} + \frac{7}{12}$ of $\frac{8}{15}$ of $(3\frac{1}{3} - \frac{45}{53}) - 5\frac{3}{4}$.

(7) What number must be added to the sum of $\frac{2}{5}, \frac{7}{8},$ **and** $1\frac{1}{2}$, to make $5\frac{89}{120}$?

(8) A bought $\frac{2}{3}$ of a cheese, and sold $\frac{1}{3}$ of his purchase to B, $\frac{1}{3}$ of what then remained to C, $\frac{1}{3}$ of what then remained to D; what part of the cheese had B, C, and D, and what part had A, after the sales?

MULTIPLICATION.

79. Rule. Multiply all the numerators together for a new numerator, and all the denominators together for a new denominator.

Ex. 1. Multiply $\frac{2}{3}$ by $\frac{5}{7}$.

By the Rule,

$$\frac{2}{3} \text{ multiplied by } \frac{5}{7} = \frac{2 \times 5}{3 \times 7} = \frac{10}{21}.$$

Reason for the Rule.
$\frac{2}{3}$ multiplied by 5, gives $\frac{10}{3}$ (Art 64.)
But $\frac{10}{3}$ must be 7 times too large, since $\frac{5}{7}$ is *one-seventh* part of 5. Therefore $\frac{10}{3}$ must be divided by 7, and $\frac{10}{3} \div 7 = \frac{10}{21}$ (Art. 65.)

Note 1. The same reasoning will apply, whatever be the number of fractions which have to be multiplied together.

Note 2. Before applying the Rule, mixed numbers must be reduced to improper fractions.

Note 3. It has been shewn that a fraction is reduced to its

VULGAR FRACTIONS.

lowest terms by dividing its num^r. and den^r. by their G. C. M., or in other words, by the product of those factors which are common to both; hence, in all cases of multiplication of fractions, it will be well to split up the num^{rs}. and den^{rs}. as much as possible into the factors which compose them; and then, after putting the several fractions under the form, of one fraction, the sign of × being placed between each of the factors in the num^r. and den^r. to cancel those factors which are common to both, before carrying into effect the final multiplication. Thus, in the following examples:

Ex. 1. Multiply $\dfrac{3}{4}$ and $\dfrac{4}{5}$ together.

Prod^t. $= \dfrac{3 \times 4}{4 \times 5} = \dfrac{3}{5}$, dividing num^r. and den^r. by 4.

Ex. 2. Multiply $\dfrac{8}{9}, \dfrac{16}{24}, \dfrac{27}{30}$, and $\dfrac{45}{60}$ together.

Prod^t. $= \dfrac{8 \times 16 \times 27 \times 45}{9 \times 24 \times 30 \times 60}$

$= \dfrac{(2 \times 2 \times 2)\times(2\times 2\times 2\times 2)\times(3\times 3\times 3)\times(3\times 3\times 5)}{(3\times 3)\times(2\times 2\times 2\times 3)\times(2\times 5\times 3)\times(2\times 2\times 3\times 5)}$

$= \dfrac{2}{5} \div$ ing by $2\times 2\times 2\times 2\times 2\times 2\times 3\times 3\times 3\times 3\times 3\times 5$.

Ex. 3. Multiplying $2\frac{1}{2}$, $3\frac{3}{8}$, $10\frac{1}{8}$, $20\frac{4}{9}$, and $5\frac{9}{23}$ together

Prod^t. $= \dfrac{5}{2} \times \dfrac{27}{8} \times \dfrac{81}{8} \times \dfrac{184}{9} \times \dfrac{124}{23}$

$= \dfrac{5\times(9\times 3)\times(9\times 9)\times(8\times 23)\times(4\times 31)}{2\times(2\times 4)\times 8\times 9\times 23}$

$= \dfrac{5\times 3\times 9\times 9\times 31}{2\times 2} = \dfrac{37665}{4} = 9416\frac{1}{4}$.

Ex. 4. Simplify ($\frac{6}{7}$ of $1\frac{1}{4}$ of $\frac{14}{15} + 3\frac{1}{2}$ of $2\frac{10}{21} - 2\frac{2}{3}$) × $3\frac{6}{7}$.

Value $= \left(\dfrac{6}{7} \text{ of } \dfrac{5}{4} \text{ of } \dfrac{14}{15} + \dfrac{7}{2} \text{ of } \dfrac{52}{21} - \dfrac{8}{3} \right) \times \dfrac{27}{7}$

$= \left(\dfrac{3\times 2\times 5\times 2\times 7}{7\times 2\times 2\times 3\times 5} + \dfrac{7\times 2\times 26}{2\times 3\times 7} - \dfrac{8}{3} \right) \times \dfrac{27}{7}$

$= \left(1 + \dfrac{26}{3} - \dfrac{8}{3} \right) \times \dfrac{27}{7} = \dfrac{3 + 26 - 8}{3} \times \dfrac{27}{7} = \dfrac{21}{3} \times \dfrac{27}{7} = 27$.

Ex. XLVI.

Find the value of

(1) $\frac{1}{2} \times \frac{3}{4}$.   (2) $\frac{7}{5} \times \frac{5}{8}$.   (3) $\frac{4}{13} \times \frac{5}{8}$.   (4) $\frac{2}{11} \times \frac{7}{8}$.

(5) $7\frac{1}{2} \times 3\frac{1}{3}$.   (6) $\frac{3}{4}$ of $\frac{1}{5} \times 17\frac{1}{2}$.   (7) $\frac{7}{12}$ of $1\frac{1}{7} \times 3\frac{3}{20} \times 1\frac{1}{2}$.

(8) $\frac{5}{8} \times 3\frac{2}{11} \times 19\frac{1}{5} \times \frac{11}{25}$.   (9) $\frac{7}{18}$ of $1\frac{1}{15}$ of $1\frac{4}{5} \times 2\frac{1}{2} \times 2\frac{2}{7}$.

(10) $1\frac{1}{6}$ of $3\frac{2}{3} \times 4\frac{1}{5}$ of $2\frac{1}{77} \times 13$.

(11) $2\frac{1}{11}$ of $(4\frac{1}{6} + 3\frac{5}{11}) \times \frac{11}{51}$ of $2\frac{1}{15} \times 1\frac{1}{10}$.

(12) $(3\frac{4}{8} - 1\frac{7}{12} + 1\frac{1}{6} - 2\frac{1}{12}) \times 38\frac{1}{4}$ of $\frac{7}{17}$.

(13) $\frac{3}{4}$ of $(\frac{1}{3} + \frac{1}{5} - \frac{1}{15} + \frac{1}{9}) \times \frac{2}{3}$ of $(2\frac{3}{10} + \frac{5}{9})$.

(14) $\{(\frac{1}{2} + \frac{1}{3})$ of $(1\frac{1}{6} + 2\frac{3}{4})\} \times \{(2\frac{1}{14} - 1\frac{1}{3})$ of $(3\frac{1}{10} - \frac{3}{7})\}$.

(15) $\{1\frac{3}{7}$ of $26\frac{1}{2}$ of $(1 - \frac{2}{3})\} \times \{2\frac{3}{5}$ of $(4\frac{1}{5} - 3\frac{3}{8})$ of $\frac{45}{106}\}$.

DIVISION.

80. Rule. Invert the divisor, *i. e.* take its numerator as a denominator and its denominator as a numerator, and proceed as in Multiplication.

Ex. 1. Divide $\frac{3}{7}$ by $\frac{2}{3}$.

By the Rule, $\frac{3}{7} \div \frac{2}{3} = \frac{3}{7} \times \frac{3}{2} = \frac{9}{14}$.

Reason for the Rule. If $\frac{3}{7}$ be divided by 2, the result is $\frac{3}{14}$ (Art. 65).

This quotient is only *one-third* part of the required quotient, since the divisor is *one-third* part of 2; hence $\frac{3}{14}$ must be multiplied by 3, in order to give the true quotient, and $\frac{3}{14} \times 3 = \frac{9}{14}$. (Art. 64.)

Note. Before applying this Rule, mixed numbers must be reduced to improper fractions, and compound fractions to simple ones.

Ex. 2. Find the quotient of $3\frac{3}{25}$ by $4\frac{2}{5}$.

$$3\frac{3}{25} \div 4\frac{2}{5} = \frac{78}{25} \div \frac{22}{5} = \frac{78}{25} \times \frac{5}{22} = \frac{2 \times 39 \times \cancel{5}}{\cancel{5} \times 5 \times 2 \times 11} = \frac{39}{55}.$$

Ex. XLVII.

Divide

(1) $1\frac{3}{20}$ by $\frac{5}{7}$.   (2) $\frac{4}{5}$ by $\frac{3}{5}$.   (3) $\frac{5}{11}$ by $\frac{3}{8}$.

(4) $4\frac{2}{5}$ by $6\frac{7}{8}$.   (5) 56 by $5\frac{9}{5}$.

(6) $7\frac{2}{9}$ by $4\frac{2}{11}$.   (7) $\frac{1}{3}$ of $20\frac{2}{3}$ by $10\frac{2}{3}$.

(8) $\frac{4}{5}$ of $5\frac{1}{2}$ by $\frac{5}{27}$ of 9.   (9) $(\frac{2}{3}$ of $7\frac{1}{2} - \frac{5}{17})$ by $1\frac{2}{5}$.

(10) Divide $\frac{1}{8} + \frac{3}{4} - \frac{1}{2}$ by the sum of $\frac{1}{8}$ and $\frac{3}{4}$.

VULGAR FRACTIONS.   87

(11) What number multiplied by 216 will produce $6\frac{3}{4}$?

(12) What must $\frac{3}{4}$ be divided by in order to produce 2?

(13) What is the least fraction which must be added to the sum of 4 and $\frac{1}{4}$ divided by their difference to make the result a whole number?

Note. COMPLEX FRACTIONS may by this Rule be reduced to simple ones.

(1) $\dfrac{1\frac{3}{4}}{2\frac{1}{2}} = \dfrac{\frac{7}{4}}{\frac{5}{2}} = \frac{7}{4} \div \frac{5}{2}$ (Art. 59) $= \frac{7}{4} \times \frac{2}{5} = \frac{7}{10}$.

(2) $\dfrac{4\frac{1}{2}}{30} = \dfrac{\frac{9}{2}}{\frac{30}{1}} = \frac{9}{2} \div \frac{30}{1} = \frac{9}{2} \times \frac{1}{30} = \frac{3}{20}$.

(3) $\dfrac{4\frac{1}{12} + 2\frac{2}{3}}{13\frac{1}{12} - 3\frac{1}{8}} = \dfrac{6 + \frac{1}{12} + \frac{2}{3}}{10 + \frac{1}{12} - \frac{1}{8}} = \dfrac{6 + \frac{1}{12} + \frac{8}{12}}{10 + \frac{1}{8} - \frac{1}{12}}$

$= \dfrac{6 + \frac{9}{12}}{10 + \frac{1}{12}} = \dfrac{\frac{81}{12}}{\frac{121}{12}} = \frac{81}{12} \times \frac{12}{121} = \frac{81}{121}$.

Simplify,   Ex. XLVIII.

(1) $\dfrac{6\frac{1}{7}}{3\frac{1}{3}}$.   (2) $\dfrac{6}{2\frac{1}{4}}$.   (3) $\dfrac{2\frac{1}{4}}{6}$.   (4) $\dfrac{6\frac{5}{7}}{3\frac{2}{3}}$.   (5) $\dfrac{5}{2\frac{5}{8}}$.

(6) $\dfrac{\frac{7}{15}}{4\frac{1}{30}}$.   (7) $\dfrac{1\frac{3}{4} \text{ of } 1\frac{1}{7}}{1\frac{2}{3} \text{ of } \frac{8}{11}}$.   (8) $\dfrac{\frac{7}{8} + \frac{3}{4}}{2\frac{1}{7}}$.   (9) $\dfrac{5\frac{1}{2} + 6\frac{3}{7}}{6\frac{3}{7} - 5\frac{1}{2}}$.   (10) $\dfrac{\frac{1}{2\frac{1}{2}} + \frac{1}{3\frac{1}{3}} + \frac{1}{4\frac{1}{4}}}{\frac{3}{6} + \frac{4}{6} - 1\frac{1}{2}}$

(11) $\left\{\dfrac{3\frac{1}{3}}{7} + \dfrac{2}{10\frac{1}{2}} - \dfrac{5}{18} \text{ of } \dfrac{4}{7}\right\} \times 1\frac{3}{4}$.   (12) $\left(\dfrac{5\frac{4}{7}}{31\frac{5}{8}} \text{ of } \frac{9}{11}\right) \div \left(\dfrac{3\frac{1}{3}}{3\frac{3}{4}} \text{ of } 15\right)$.

(13) $\dfrac{5\frac{3}{7} \div 7\frac{2}{6}}{2\frac{3}{5} \cdot 1\frac{4}{7}} \text{ of } \dfrac{2\frac{1}{4} \times 8\frac{1}{3}}{4\frac{1}{6} \div (\frac{1}{3} - \frac{2}{9})}$.   (14) $\dfrac{13}{2\frac{2}{3} + \frac{1}{3}} + \dfrac{1\frac{1}{3}}{3\frac{1}{5}} - 1\frac{3}{13}$.

81. *To find the* **value of** *a fraction* **in** *terms of the same or lower denomination.*

RULE. Divide (if possible) the numerator by the denominator; if there be a remainder, reduce it to the next lower name, and divide the product by the denominator; repeat the latter operation as often as necessary.

Find the value of $\frac{2}{7}$ of £15.

By the Rule,

$\frac{2}{7}$ of £15 $= £\dfrac{2 \times 15}{7} = £\dfrac{30}{7} = £4\frac{2}{7}$; $£\frac{2}{7} = \dfrac{2 \times 20}{7}s. = \dfrac{40}{7}s. = 5\frac{5}{7}s.$ &

ARITHMETIC

$$\tfrac{4}{7}s. = \frac{5 \times 12}{7} d. = \frac{60}{7} d. = 8\tfrac{4}{7}d.; \quad \tfrac{4}{7}d. = \frac{4 \times 4}{7}q. = \frac{16}{7}q. = 2\tfrac{2}{7}q.$$

$$\therefore \tfrac{4}{7} \text{ of } £15 = £4.\ 5s.\ 8\tfrac{1}{2}d.\ \tfrac{2}{7}q.$$

Ex. XLIX.

Find the respective values of,

(1) $\tfrac{3}{4}$ of $1. (2) $\tfrac{3}{8}$ of a ml. (3) $\tfrac{4}{9}$ of a cwt.
(4) $\tfrac{9}{10}$ of 2 tons. 3 cwt. (5) $\tfrac{3}{10}$ of 3 mls., 2 fur.
(6) $\tfrac{4}{5}$ of 3 ac., 2 per., 3 yds. (7) $\tfrac{5}{8}$ of 5 lbs., 13 dwts.
(8) $\tfrac{7}{8}$ of 63 yds. 2 nls. (9) $\tfrac{3}{11}$ of £26. 8s. 11d.
(10) $\tfrac{4}{7}$ of 128 lbs., 2 sc. (11) $\tfrac{7}{8}$ of $\tfrac{3}{8}$ of $10\tfrac{2}{3}$ hrs.
(12) $7\tfrac{2}{3}$ of a lb. Avoird. (13) $\tfrac{4}{7}$ of $\tfrac{2}{3}$ of $42.
(14) $\tfrac{9}{10}$ of a day. (15) $\tfrac{9}{13}$ of 24 cords of wood.

82. *To reduce a given quantity to the fraction of another quantity of the same kind.*

RULE. Reduce both to the same name; and take the result of the former for the numerator, and of the latter for the denominator, of the required fraction.

Reduce 7s. 5d. to the fraction of £1.

Method of working.

7s. 5d. = 89d.
£1. = 240d.
∴ the fraction is $\tfrac{89}{240}$.

Reason for the Rule.

For 1d. = $\tfrac{1}{240}$ of £1; ∴ 7s. 5d. which = 89d. is $\tfrac{89}{240}$ of £1.

Ex. L.

Reduce,

(1) 3s. 4d. to the fr. of £1.
(2) 2 ro. 13 per. to the fr. of 3 acres.
(3) 3 wks., 16 min. to the fr. of half-an-hour
(4) 1 lb., 1 oz., 3 dwt., to the fr. of 2 lbs.
(5) 1 lb., 5 oz. to the fr. of 2 lbs. 1 sc.
(6) 8 ac., 3 ro. to the fr. of 2 ac., 32 per.
(7) 2 sq. yds., 2 ft., 120 in., to fr. of 3 per. $13\tfrac{1}{4}$ yds., 1 ft., 72 in.
(8) £1. 18s. to the fr. of £7.
(9) 2 bu., 1 pk., to the fr. of 4 bu. 1 gal.
(10) $2.09 to the fr. of $56.43.
(11) 2 yds., 2 ft. to the fr. of 13 per., 3 yds., 6 in.

VULGAR FRACTIONS.

(12) 1 lb. Troy to the fr. of 1 lb. Avoirdupois.
(13) What fraction of 7 bu. is 3 qts.?
(14) What fraction of 4 mls., 2 fur. is 1½ yds.?
(15) What fraction of 5 ac., 1 per. is 1 yd., 4 in.?

83. *To reduce a fraction of one given quantity to a fraction of another.*

RULE. Express by (82) the first quantity as a fraction of the second; and the fraction required will then be found by reducing the resulting compound fraction to a simple one.

Ex. 1. Reduce $\frac{2}{7}$ lb. to the fraction of a cwt.

Method of working,

1 lb. $= \frac{1}{100}$ of cwt.; $\therefore \frac{2}{7}$ lb. $= \frac{2}{7} \times \frac{1}{100}$ of a cwt. $= \frac{2}{700}$ of owt.

Ex. 2. $2\frac{1}{4}$ of $5.25 to the fraction of 15 cents.

$5.25 is $\frac{35}{100}$ of 15 cts.; $\therefore 2\frac{1}{4}$ of $\frac{35}{100}$ of 15 cts. $= \frac{315}{400}$ of 15 cts.

Ex. LI.

Reduce,

(1) $\frac{2}{7}$ of $14 to the fr. of $\frac{1}{2}$ of $16.
(2) $\frac{5}{8}$ of 2 ac., 2 ro. to the fr. of $\frac{1}{3}$ of 3 ac., 2 per.
(3) $2\frac{1}{2}$ of 3 lbs., 6 dwt. to the fr. of $1\frac{1}{3}$ of 6 lbs., 12 grs.
(4) $12\frac{2}{3}$ of 3s. 6d. to the fr. of £1.
(5) $3\frac{1}{3}$ of 10 cwt., 2 qrs., to the fr. of 1 ton.
(6) $3\frac{1}{2}$ of 2 ac., 3 ro. to the fr. of 2 ro., $2\frac{1}{2}$ per.
(7) $\frac{6}{8}$ lb. Troy to the fr. of a lb. Av.
(8) $1\frac{3}{17}$ of £2. 4s. $7\frac{1}{2}d$. to the fr. of 5s.
(9) $\frac{1}{11}$ of $2\frac{3}{4}$ mls. to the fr. of $\frac{1}{2}$ of $\frac{7}{8}$ mls.
(10) $6\frac{1}{2}$ of 3 cords to the fr. of 5 cord ft.
(11) $8\frac{1}{3}$ of 6 lbs., 2 sc. to the fr. of a lb.
(12) $\frac{4}{5}$ of $\frac{3}{8}$ of $21 to the fr. of $7.
(13) $\frac{7}{12}$ of 8 yds., 2 nls. to the fr. of $2\frac{1}{2}$ ells (English).
(14) $2\frac{2}{3}$ of 10 hrs. to the fr. of 1 day.

84. *Miscellaneous Examples in Vulgar Fractions worked out.*

Ex. 1. At the 'call over' at a certain school, $\frac{5}{8}$ of the children on the register answered to their names; the rest, 15 in number, were absent. How many children were there on the register?

ARITHMETIC.

$\frac{5}{6}$ of the no. were present, ∴ $\frac{1}{6}$ of no. were absent.
By the question, $\frac{1}{6}$ of no. = 18.
∴ no. = 18 × 6 = 108.

Ex. 2. A poor woman lost through a hole in her pocket $\frac{4}{11}$ of her money; only 3s. 0¾d. was left. How much money had she at first, and how much did she lose?

After losing $\frac{4}{11}$ of her money, $\frac{7}{11}$ of it was left,

∴ $\frac{7}{11}$ of her money = 3s. 0¾d.
∴ $\frac{1}{11}$ of her money = 3s. 0¾d. ÷ 7 = 5¼d.
∴ her money = 5¼d. × 11 = 4s. 9¾d.

She lost $\frac{4}{11}$ of 4s. 9¾d. = $\dfrac{19s.\ 3d.}{11}$ = 1s. 9d.

Ex. 3. A, B, C, D run a race over 1 mile. First A and B race, when A wins by 20 yds.; then C and D race, when C wins by 60 yds.; then A and C race, which will win, and by how much, supposing that if B and D had run against each other, B would have won by 40 yds.?

While A runs 1760 yds., B runs 1740 yds.; while C runs 1760 yds., D runs 1700 yds., or while D runs 1 yd., C runs $1\frac{7\ 3\ 0}{1\ 7\ 0\ 0}$ yds.; while B runs 1760 yds, D would have run 1720 yds., or while B runs 1 yd., D would have run $\frac{1720}{1760}$ yds.
While A runs 1760 yds., B runs 1740 yds.
" A " " D runs (1740 × $\frac{172}{176}$) yds.
" A " " C runs (1740 × $\frac{172}{176}$ × $\frac{17}{170}$), or 1760 $\frac{8}{17}$ yds.

∴ C will win by $\frac{8}{17}$ yds.

Ex. 4. Divide 15s. 6d. between A and B, so that B's share may be less than A's share by $\frac{2}{5}$ of A's share.

To represent A's share fix on some number which is exactly divisible by 5; let 5 represent A's share.
Then B's share = 5 − $\frac{2}{5}$ of 5, or 5 − 2, or 3.
∴ 15s. 6d. has to be divided into 5 + 3, or 8 shares, of which A is to have 5, and B 3;

∴ value of each share = $\dfrac{15s.\ 6d.}{8}$ = 1s. 11¼d.

∴ A's share = 1s. 11¼d. × 5 = 9s. 8¼d., B's = 1s. 11¼d. × 3 = 5s. 9¾d.

Ex. 5. If 7 men or 11 boys can dig a field in 10 days, in what time will 11 men and 7 boys dig a field of half the size?

7 men = 11 boys, ∴ 1 man = $\frac{11}{7}$ boy;
∴ 11 men and 7 boys = (11 × $\frac{11}{7}$ + 7), or $\dfrac{121 + 49}{7}$, or $\dfrac{170}{7}$ boys.

VULGAR FRACTIONS.

By the question,
11 boys can dig the greater field in 10 days,
$\therefore$ 1 boy........................(10×11) days;
$\therefore \frac{170}{7}$ boys......................$\frac{10 \times 11 \times 7}{170}$ days;

$\therefore$ .... the less field in $\frac{10 \times 11 \times 7}{170 \times 2}$ days $= 2\frac{9}{34}$ days.

Ex. 6. Divide 1860 cords of wood between A, B, and C, so that for every 5 cords given to A, B may receive 4 cords, and for every 3 cords given to B, C may receive 1 cord.

The L. C. M. of 5, 4, and 3 is 60; $\therefore$ if 60 shares be given to A, B will have $\frac{4}{5}$ of 60 shares, or 48 shares, and C will have $\frac{1}{3}$ of 48 shares, or 16 shares;

$\therefore A$, B, and C together have $(60 + 48 + 16)$, or 124 shares;

$\therefore A$ has $\frac{60}{124}$ of 1860 cords $= (15 \times 60)$, or 900 cords.

B has $\frac{48}{124}$ of 1860 cords $= (12 \times 60)$, or 720 cords.

C has $\frac{16}{124}$ of 1860 cords $= (4 \times 60)$, or 240 cords.

Ex. 7. A can do a piece of work in 5 days, B can do it in 6 days, and C can do it in 7 days; in what time will A, B, and C, all working at it, finish the work? Find also in what time A and B working together, A and C together, and B and C together, could respectively finish it.

In one day....A....does $\frac{1}{5}$ part of the work,
..............B........$\frac{1}{6}$...............
..............C........$\frac{1}{7}$...............,

$\therefore$ ....$A + B + C$ do $\left(\frac{1}{5} + \frac{1}{6} + \frac{1}{7}\right)$ or $\frac{107}{210}$;

$\therefore$ no. of days in which $A + B + C$ would finish the work
$= \dfrac{\text{whole work}}{\text{part done in one day}} = \dfrac{1}{\frac{107}{210}} = \dfrac{210}{107} = 1\frac{103}{107}$.

Again, in one day $A + B$ do $\left(\frac{1}{5} + \frac{1}{6}\right)$, or $\frac{11}{30}$ of the work,

$\therefore A + B$ would finish the work in $\dfrac{1}{\frac{11}{30}}$, or $\frac{30}{11}$, or $2\frac{8}{11}$ days.

In like manner, it may be shewn that A and C would finish the work in $2\frac{1}{2}$ days; and B and C in $3\frac{1}{15}$ days.

Ex. LII.

(1) $\frac{4}{5}$ths of a farm belongs to A, the rest to B; A sells $\frac{2}{3}$ths of his share to C, and $\frac{1}{17}$th of it to B; what portions of the farm do A, B, and C, respectively hold after the sales?

(2) (1) Among how many boys can 9 oranges be divided, so that each boy may have $\frac{3}{4}$ of an orange? (2) From the sum of $4\frac{1}{2}$ and $3\frac{9}{10}$ take their difference.

(3) Divide $\frac{7}{8}$ into two parts, so that one of them is greater than the other by $\frac{3}{8}$.

(4) (1) What number must be multiplied by $1\frac{1}{2}$ of $2\frac{3}{4}$ to give $3\frac{2}{7}$? (2) What number must be added to $\frac{1}{2}$ of $2\frac{1}{3}$ to give $3\frac{2}{7}$?

(5) A gives to B $\frac{1}{3}$ of his money, to C $\frac{1}{2}$ of what remains, and to D $\frac{1}{3}$ of what then remains; compare the sums which A and D will now have.

(6) Miss Taylor, after spending $\frac{1}{3}$rd of the money in her purse, and then $\frac{2}{3}$ths of the remainder, has still left $4.20; how much had she in her purse at first?

(7) $\frac{3}{13}$ of a fishing smack being worth $90, find the value of $\frac{2}{3}$rd of it.

(8) A person after paying an income-tax of 5 cents in the dollar, has a net income of $855; find his gross income?

(9) If, when the income-tax was 6 cents in the dollar, a person paid $54; how much less will he now pay, the tax being reduced 4 cents in the $?

(10) If $\frac{4}{5}$ of a rabbit be worth $\frac{3}{8}$s., and $\frac{1}{8}$ of a rabbit be worth $\frac{1}{50}$ of a pig; what is the value of 100 pigs?

(11) If, in practising, 7 riflemen shoot 26 rounds in 1 hr., 31 min.; how many rounds will 37 riflemen shoot in $4\frac{1}{2}$ hrs. at the same rate?

(12) A sum of money is divided into 4 parts, which are to each other as the numbers 1, 2, 3, 4; and a person, who receives $\frac{3}{4}$ of each share, obtains altogether $12.60; find the sum of the several shares?

(13) If 15 cows or 28 sheep can graze a field of 5 ac. in 11 days, how many days ought a similar field of 18 ac. to serve 33 cows and 20 sheep?

VULGAR FRACTIONS.

(14) Divide $94.50 between A and B; (1) giving A half as much again as B; (2) giving A's share less half A's share to B.

(15) A bankrupt owes to one creditor 500 dollars, to each of two others $250, to each of three others $75: his property is worth $625. How much can he pay in the dollar, and how much will the first creditor receive?

(16) A mine is worth $10000; a person for $\frac{3}{8}$ of his share receives $750. What part of the mine did he possess?

(17) A school is composed of three divisions; there are $\frac{4}{9}$ths of the whole number of boys in the first, $\frac{1}{4}$th in the second, and the rest, 80 in number, in the third; how many boys are there altogether?

(18) A can do a piece of work in 10 days, which B could do in 12; in what time would they do it together?

(19) A father left to the elder of his two sons $\frac{4}{5}$ of his estate, and $\frac{1}{3}$ of the remainder to the younger, and the residue to the widow; find their respective shares, it being found that the elder son received $1690 more than the younger.

(20) Divide 85 ac. 2 ro. of land between A, B, and C, so that B's share $= \frac{6}{7}$ of A's share, and that C's share shall be 9 ac. more than the united shares of A and B.

(21) A fine of $14.40 had to be raised among a number of boys; one-third paid 18 cents each, as many more 30 cents each, and the remainder 42 cents each. How many boys were there?

(22) A cistern has 3 pipes in it, by one of which it could be filled in 3 minutes, and by the other two it could be emptied in 6 and 7 minutes respectively; in what time will it be filled, if they are all opened together?

(23) A and B together can do a piece of work in 30 days, B by himself can do it in 70 days; (1) in what time could A do it by himself? (2) how much more of the work does A do than B, when they work together?

(24) A and B can do a piece of work in $6\frac{2}{3}$ days, A and C in $5\frac{1}{2}$ days, and A, B, and C in $3\frac{3}{4}$ days. In how many days can A do it alone?

(25) There are 4 casks of different sizes. The 1st is filled with liquid the rest are empty. The 2nd cask is filled from the 1st, and $\frac{2}{5}$ths of the original liquid in the 1st remains. The 3rd is then filled from the 2nd, and $\frac{1}{4}$th of the liquid in

94   ARITHMETIC.

the 2nd remains. The liquid in the third is then poured into the 4th, and fills $\frac{9}{10}$ths of it. Had the 3rd and 4th casks been filled from the contents of the 1st, 15 gallons would still have remained in the 1st. Find the size of each cask?

(26) A in 2 days can do as much work as B can do in 3 days; together they take 12 days to do a certain work. In what time would A alone have done it?

DECIMALS.

85. Figures in the units' place of any number express their *simple* values, while those to the *left* of the units' place increase in value *tenfold* at each step from the units' place; therefore, according to the same notation, as we proceed from the units' place to the *right* every successive figure would decrease in value *tenfold*. We can thus represent whole numbers or integers and certain fractions under a uniform notation by means of figures in the units' place and on each side of it; for instance, in the number 5673·241, the figures on the left of the *dot* represent *integers*, while those on the right of the dot denote *fractions*. The number written at length would stand thus:

$$5 \times 1000 + 6 \times 100 + 7 \times 10 + 3 + \frac{2}{10} + \frac{4}{100} + \frac{1}{1000}.$$

The dot is termed the decimal point, and all figures to the right of it are called DECIMALS, or DECIMAL FRACTIONS, because they are fractions with either 10, 100 or 10 × 10, 1000 or 10 × 10 × 10, &c., as their respective denominators.

The *extended Numeration Table* will be represented thus:

7	6	5	4	3	2	1 .	2	3	4	5	6	7	
&c.	Millions.	Hundreds of Thousands.	Tens of Thousands.	Thousands.	Hundreds.	Tens.	Units.	Tenths.	Hundredths.	Thousandths.	Ten Thousandths.	Hundred Thousandths.	Millionths.	&c.

86. 10, called the *first* POWER of 10, is written thus, 10^1.

10×10, or 100, called the *second* POWER of 10, is written thus, 10^2.

$10 \times 10 \times 10$, or 1000, called the *third* POWER of 10, is written thus, 10^3, and so on; similarly of other numbers: thus the fifth power of 4 is $4 \times 4 \times 4 \times 4 \times 4$, and is written thus, 4^5.

The small figures 1, 2, 3, &c., at the right of the number, a little above the line, are called INDICES.

87. $\cdot 306 = \dfrac{3}{10} + \dfrac{0}{100} + \dfrac{6}{1000} = \dfrac{3 \times 100}{10 \times 100} + \dfrac{0 \times 10}{100 \times 10} + \dfrac{6}{1000}$

$= \dfrac{300}{1000} + \dfrac{0}{1000} + \dfrac{6}{1000} = \dfrac{306}{1000}.$

Again, $\cdot 0306 = \dfrac{0}{10} + \dfrac{3}{100} + \dfrac{0}{1000} + \dfrac{6}{10000} = \dfrac{0 \times 1000}{10 \times 1000}$

$+ \dfrac{3 \times 100}{100 \times 100} + \dfrac{0 \times 10}{1000 \times 10} + \dfrac{6}{10000} = \dfrac{0 + 300 + 0 + 6}{10000} = \dfrac{306}{10000}.$

Again, $80 \cdot 306 = 80 + \dfrac{306}{1000} = \dfrac{80000 + 306}{1000} = \dfrac{80306}{1000}.$

Hence to convert decimals to vulgar fractions: from the above examples we deduce the following:

88. RULE. Write the figures which compose the decimal as numerator, and for denominator 1, followed by as many cyphers as there are figures after the decimal point.

Ex. LIII.

Express as vulgar fractions.

(1) $\cdot 3$; $\cdot 13$; $\cdot 19$; $\cdot 301$; $\cdot 270$; $\cdot 5653$.

(2) $\cdot 504$; $\cdot 73201$; $\cdot 791003$; $\cdot 03$; $\cdot 0045$.

(3) $\cdot 300$; $18 \cdot 741$; $2 \cdot 1$; $\cdot 00001$; $5 \cdot 0007$.

(4) $347 \cdot 02007$; $500 \cdot 005$; $5 \cdot 60746803$; $\cdot 0000500$.

(5) $29 \cdot 0050$; $20 \cdot 607$; $5 \cdot 00058$.

89. Any fraction, having 10, or any power of 10, for its denominator, as $\frac{80036}{10000}$, may be expressed thus, $80 \cdot 0036$.

For $\frac{80036}{10000} = 80 + \frac{3}{1000} + \frac{6}{10000} = 80 + \frac{0}{10} + \frac{0}{100} + \frac{3}{1000} + \frac{6}{10000}$

$= 80 \cdot 0036$ (by the Notation we have assumed).

90. $\cdot 241 = \frac{241}{1000}$, $\cdot 0241 = \frac{241}{10000}$, $\cdot 2410 = \frac{2410}{10000} = \frac{241}{1000}$.

We see that $\cdot 241$, $\cdot 0241$, and $\cdot 2410$ are respectively equivalent to fractions which have the same numerator, and the first and third of which have also the same denominator, while the denominator of the second is greater. Hence ·241 is equal to ·2410, but ·0241 is less than either.

The value of a decimal is therefore not affected by *affixing* cyphers to the right of it; but its value is decreased by *prefixing* cyphers: which effect is exactly opposite to that which is produced by affixing and prefixing cyphers to integers.

91. A decimal is *multiplied* by 10, if the decimal point be removed *one* place towards the *right* hand; by 100, if *two* places; by 1000, if *three* places; and so on: and conversely, a decimal is *divided* by 10, if the point be removed *one* place to the *left* hand; by 100, if *two* places; by 1000, it *three* places; and so on.

Thus, $5·6 \times 10 = \frac{56}{10} \times 10 = 56$; $5·6 \times 1000 = \frac{56}{10} \times 1000 = 5600$.
$5·6 \div 10 = \frac{56}{10} \times \frac{1}{10} = \frac{56}{100} = ·56$; $5·6 \div 1000 = \frac{56}{10} \times \frac{1}{1000} = \frac{56}{10000}$
$= ·0056$.

Ex. LIV.

(1) $\frac{4}{10}$; $\frac{28}{10}$; $\frac{233}{10}$; $\frac{14}{100}$; $\frac{147}{1000}$; $\frac{17}{1000}$.

(2) $\frac{5083}{10}$; $\frac{941}{100}$; $\frac{26}{100000}$; $\frac{503}{100}$; $\frac{503}{100000}$.

(3) $\frac{55600}{1000}$; $\frac{1700701}{100000}$; $\frac{50005}{10000000}$; $\frac{2}{10000000}$; $\frac{2078854}{100000}$; $\frac{53053}{10000000000}$.

(4) Seven-tenths; thirty thousandths.

(5) Three hundred and three thousandths; one ten thousandth.

(6) Four, and five hundred and four millionths; seventy ten millionths.

Express in words the meaning of,

(7) ·6; ·17; ·07.   (8) ·007; ·700; 6·3004.

(9) 35·00205;   400·34000.

(10) Multiply ·3, ·13, ·013, 54·0003, 7420·1, each separately by 10, 100, 10000, and by ten millions.

(11) Divide 5·362, ·3, 70·0107, and 5000, each separately by 10, 100, and by 1000000.

(12) What is the quotient of 2·03 by a million?

ADDITION OF DECIMALS.

92. RULE. Place the numbers under each other, units under units, tens under tens, &c., tenths under tenths, &c.; so that the decimals be all under each other. Add as in

DECIMALS. 97

whole numbers, **and place the decimal point in the sum under** the decimal point above.

Ex. Add together 2·3, ·056, 37, and 3·60015.

By the Rule.

2·3
·056
37·
3·60015
─────
42·95615

By fractions.

$2·3 + ·056 + 37 + 3·60015 = \frac{23}{10} + \frac{56}{1000} + \frac{37}{1} + \frac{360015}{100000}$
$= \frac{230000}{100000} + \frac{5600}{100000} + \frac{3700000}{100000} + \frac{360015}{100000}$
$= \frac{4295615}{100000} =$ 42·95615 (Art. 89).

Ex. LV.

Add

	(1)	(2)	(3)	(4)
	1·035	24·	186·8	94·25
	·00643	185·3009	35·2779	·008
	27·	·98795	9000·	187·96009
	2·2146	3·098	9·291	57·3916
	530·09	·70006	830·05764	5·998347

Add together, and verify each result by fractions :

(5) 12·5, 20·043, 7·63201, add ·0561.

(6) ·0573, 15, 2·04, and 567·98075.

(7) 503·0003, 13·98, 5853·097, and 960.

(8) 6·00734, 54, 15·70087012, 8·00003, and 9·987789.

(9) Find the sum of thirteen hundredths, seven and three ten-thousandths, four hundred and eight and five tenths, nine hundred and seventy-eight, and eight hundred and eight ten-thousandths.

SUBTRACTION OF DECIMALS.

93. RULE. Place the less number under the greater, units under units, tens under tens, &c., tenths under tenths, &c.; suppose cyphers to be supplied if necessary in the upper line to the right of the decimal.

Then subtract as in whole numbers, and place the decimal point in the remainder under the decimal point above.

Ex. Subtract 3·084 from 5·7.

By the Rule,

5·7
3·084
─────
2·616

By fractions,

$5·7 - 3·084 = \frac{57}{10} - \frac{3084}{1000} = \frac{5700}{1000} - \frac{3084}{1000}$
$= \frac{2616}{1000} =$ 2·616.

7

Ex. LVI.

(1) From 5·345 (2) 26·002 (3) 15·67 (4) 21
    Take 3·087   18·9564   9·7003   19·9009

(5) Find the difference between, verifying each result by fractions, (1) ·13 and 1·3 ; 2·07 and 207. (2) 76·3 and 7.63 ; 67·3 and 67·5803. (3) 501 and 428·90456 ; 53·24 and 5324. (4) 4·42 and ·00042 ; ·0000007 and ·007.

(6) By how much does 23 exceed the difference between 2·3 and ·23 ?

(7) Find the difference (1) between one-tenth and five thousandths; (2) between twenty and nine thousandths and twenty-nine thousandths.

(8) A person who has seven-tenths of a ship, sells eighty-seven thousandths of it, how much has he left ?

(9) Find the least fraction, which added to the sum of 1·2, ·12, ·012, and 210, will make the result a whole number.

(10) Find the value of (1) $1·25 — 3·059 + 235·6758 — 184·0003 ; (2) 215·263 — (7·0004 — ·05) — (45·08 + 80·5007).

MULTIPLICATION OF DECIMALS.

94. Rule. Multiply the numbers together as if they were whole numbers, and point off in the product as many decimal places as there are decimal places in both the multiplicand and the multiplier ; if there are not figures enough, supply the deficiency by prefixing cyphers.

Ex. Find the product of (1) 7·35 by ·23, (2) of 8·27 by ·0002.

By the Rule,
(1)   7·35
      ·23
   ——
  2205
  1470
  ————
  1·6905

By fractions,
$7·35 \times ·23 = \frac{735}{100} \times \frac{23}{100} = \frac{16905}{10000} = 1·6905$.

(2)   8·27
    ·0002
  ————
  ·001654

$8·27 \times ·0002 = \frac{827}{100} \times \frac{2}{10000} = \frac{1654}{1000000} = ·001654$.

Ex. LVII.

Multiply
(1) 3·25 (2) 6·035 (3) 40·004 (4) 680·35 (5) 20607
By ·35   2·7   2·03   ·0049   ·20607

DECIMALS. 99

Multiply, and verify **each result by fractions**:

(6) 60·71 by 11; 57·068 by 2·004; 5·36 by 700; 7·01509 by 50·805.

(7) 48·067 by ·00037; 54·3047 by 9·00005; 2·568 by ·00025.

(8) Find the **continued product** (1) of 5·5, ·055, 550, and ·0055; (2) of 1·75, 6·2, 85, and ·0004.

(9) How many yds. of cloth are there in 7·35 pieces of cloth, each of which contains 37·85 yds.?

(10) A man eats ·95 of a loaf daily; how many loaves will he eat in the year 1866?

DIVISION OF DECIMALS.

95. *First.* *When the number of decimal places in the dividend exceeds the number of decimal places in the divisor.*

RULE. Divide as in whole numbers, and mark off in the quotient a number of decimal places equal to the excess of the number of decimal places in the dividend over the number of decimal places in the divisor; if there are not figures sufficient, prefix cyphers as in Multiplication.

Ex. 1. Divide (1) 2·1125 by 8·45, (2) ·0021125 by 84·5.

By the Rule,

(1)   8·45) 2·1125 (25
          1690
          ─────
          4225
          4225

By fractions,

$$2\cdot1125 \div 8\cdot45 = \tfrac{21125}{10000} \div \tfrac{845}{100} = \tfrac{21125}{10000} \times \tfrac{100}{845} = \tfrac{21125}{845} \times \tfrac{1}{100} = \tfrac{25}{1} \times \tfrac{1}{100} = \tfrac{25}{100} = \cdot 25.$$

No. of decl. places in quotient $= 4 - 2 = 2$, ∴ quotient $= \cdot 25$.

(2)   84·5) ·0021125 (25
           1690
           ─────
           4225
           4225

By fractions,

$$\cdot 0021125 \div 84\cdot5 = \tfrac{21125}{10000000} \div \tfrac{845}{10} = \tfrac{21125}{10000000} \times \tfrac{10}{845} = \tfrac{21125}{845} \times \tfrac{1}{1000000} = \tfrac{25}{1} \times \tfrac{1}{1000000} = \tfrac{25}{1000000} = \cdot 000025.$$

No. of decl. places in quotient $= 7 - 1 = 6$, ∴ quotient $= \cdot 000025$.

96. *Secondly.* *When the number of decimal places in the dividend is less than the number of decimal places in the divisor.*

RULE. Affix cyphers to the dividend until the number of decimal places in the dividend equals the number of decimal places in the divisor; the quotient up to this point of the division will be a whole number.

If there be a remainder, and the division be carried on

further, the figures in the quotient after this point will be decimals.

Ex. 2. Divide 2112·5 by ·845.

By the Rule,

```
·845)2112·500(2500
     1690
     ────
     4225
     4225
     ────
       00
```

By fractions,

$2112 \cdot 5 \div \cdot 845 = \frac{21125}{10} \div \frac{845}{1000} = \frac{21125}{10} \times \frac{1000}{845} = \frac{21125}{845} \times \frac{1000}{10} = 25 \times 100 = 2500.$

Ex. LVIII.

Divide and verify each result by fractions.

(1) 33·372 by 2·7.   (2) ·33372 by ·27.   (3) ·33372 by 27.
(4) 33372 by ·27.   (5) 33372 by ·00027.
(6) 561·0833 by ·323.   (7) 5610833 by ·323.
(8) 56108·33 by 3·23.   (9) 5610·833 by ·0000323.
(10) 552·5325 by 3·25, and also by ·00325.
(11) 2·419003 by 464·3, and also by ·004643.
(12) ·000081 by 2·7, by ·0027, and also by 27000.
(13) 218051·081884 by 2·00099, and by 200099.
(14) ·121 by 11, by 1100, and also by ·0011.
(15) 393·72 by ·000193, by 1·93, and also by 193000.
(16) 590·4825 by ·03275, and also by 327500.
(17) 213·419596 by 1·00103, and also by 100103.
(18) Divide the sum of twenty-four ten thousandths and twenty-four hundredths by twenty-four.
(19) Two ten thousandths by twenty-five hundredths.
(20) If a man mow 1·75 ac. of grass in one day, how long will it take him to mow 21·875 ac.?
(21) How often is ·75 min. contained in 64·125 min.?
(22) The product of two numbers is seventy thousand two hundred and forty two hundred millionths; one of the numbers is twenty-three thousandths; find the other number.

Ex. 3. Divide 240·13 by 73·4 to three places of decimals.

Before dividing, affix two cyphers to the dividend, so as to make the number of decimal places in the dividend exceed

DECIMALS.

the number of decimal places in the divisor by 3; if we divide up to this point, the quotient will contain 3 decimal places by Rule 1.

```
73·4 ) 240·1300 ( 3·271
       220
       ───
       1993
       1468
       ────
       5250
       5138
       ────
       1120
        734
       ────
        386
```

By fractions,

$240 \cdot 13 \div 73 \cdot 4 = \frac{24013}{100} \div \frac{734}{10}$

$= \frac{24013}{100} \times \frac{10}{734} = \frac{24013}{734} \times \frac{10}{100}$

$= \frac{24013}{734} \times \frac{100}{1000}$.

(we multiply numr. and denr. by 10, to make denr. 1000, since the quotient is to contain *three* decimal places)

$= \frac{2401300}{734} \times \frac{1}{1000} = \frac{3271}{1000} = 3 \cdot 271$.

Ex. LIX.

Divide to three places of decimals, and verify each result by fractions,

(1) 1·9 by ·3, by ·03, and by 300.

(2) 4·937 by 159, by 1·59, and by 1590.

(3) 329744 by 53, by ·0053, and by 5300.

97. *Certain Vulgar Fractions can be expressed accurately as Decimals.*

RULE. Reduce the fraction to its lowest terms; then place a dot after the numerator and affix cyphers for decimals; divide by the denominator, as in division of decimals, and the quotient will be the decimal required.

Ex. 1. Convert $\frac{3}{4}$, $\frac{3}{40}$, $\frac{3}{400}$, each into a decimal.

```
4 ) 3·00
    ────
     ·75
```

No. of decl. places in quotient = no. of decl. places in dividend − no. of decl. places in divisor = 2 − 0 = 2.

$\frac{3}{40} = \frac{3}{4} \div 10 = \cdot75 \div 10 = \cdot075$; $\frac{3}{400} = \frac{3}{4} \div 100 = \cdot75 \div 100 = \cdot0075$.

Ex. 2. Reduce $\frac{5}{16}$ to a decimal.

```
16 ) 5·0000 ( ·3125
     48
     ──
      20
      16
      ──
      40
      32
      ──
       80
       80
       ──
```

or thus, $16 \begin{cases} 4 & 5 \cdot 00 \\ 4 & \overline{1 \cdot 2500} \\ & \cdot 3125 \end{cases}$

$\therefore \frac{5}{16} = \cdot 3125$

Ex. 3. Convert $5\frac{3}{640} + \cdot75$ of $\dfrac{6}{5}$ of $7\frac{1}{2}$ into a decimal.

$$640 \begin{cases} 8 \\ 8 \\ 10 \end{cases} \begin{array}{|l} 5\cdot000 \\ \hline \cdot625000 \\ \hline \cdot078125 \\ \hline \cdot0078125 \end{array} \qquad \cdot75 \text{ of } \frac{6}{5} \text{ of } 7\tfrac{1}{2} = \cdot75 \text{ of } \tfrac{6}{5} \text{ of } \tfrac{15}{2}$$

$$= \cdot75 \times 9 = 6\cdot75.$$

$\therefore 5\frac{3}{640} + \cdot75$ of $\dfrac{6}{5}$ of $7\tfrac{1}{2} = 5\cdot0078125 + 6\cdot75 = 11\cdot7578125$.

Ex. LX.

Reduce to decimals,

(1) $\tfrac{1}{4}$; $\tfrac{3}{8}$; $\tfrac{4}{9}$; $6\tfrac{1}{2}$; $\tfrac{39}{8}$; $\tfrac{5}{9}$; $5\tfrac{7}{10}$. (2) $\tfrac{3}{13}$; $8\tfrac{14}{15}$; $\tfrac{19}{20}$; $\tfrac{31}{32}$; $7\tfrac{21}{25}$.

(3) $\tfrac{47}{60}$; $4\tfrac{7}{125}$; $\tfrac{3}{500}$; $\tfrac{99}{625}$; $84\tfrac{13}{1024}$.

(4) $\tfrac{5}{8}$ of $1\tfrac{3}{10}$; $3\tfrac{1}{2}$ of $2\tfrac{4}{5}$; $3\tfrac{1}{2}$ of $4\tfrac{1}{4}$ of $5\tfrac{1}{2}$.

(5) $1\tfrac{3}{5} - 1\tfrac{3}{10} + 3\tfrac{5}{10}$; $11\tfrac{1}{2} + \cdot75$ of $\tfrac{24}{25}$ of $6\tfrac{3}{4}$.

98. To convert a vulgar fraction into a decimal, we have in fact, after reducing the fraction to its lowest terms, and affixing cyphers to the num^r., to divide 10, or some multiple of 10 or of its powers, by the den^r.; now $10 = 2 \times 5$, and these are the only factors into which 10 can be broken up; therefore, when the fraction is in its lowest terms, if the den^r. be not composed solely of the factors 2 and 5, or one of them, or of powers of 2 and 5, or one of them, then the division of the num^r. by the den^r. will never terminate. Decimals of this kind are called indeterminate decimals, and they are also called Circulating, Repeating, or Recurring Decimals, from the fact that when a decimal does not terminate, the same figures must come round again, or recur, or be repeated: for since we always affix a cypher to the dividend, whenever any former remainder recurs, the quotient will also recur.

99. Pure Circulating Decimals are those which recur from the beginning: thus, $\cdot333..$, $\cdot2727..$, are pure circulat^g. dec^{ls}.

Mixed Circulating Decimals are those which do not begin to recur, till after a certain number of figures: thus, $\cdot128888..$, $\cdot0113636..$, are mixed circulat^g. dec^{ls}.

The circulating part is called the Period or Repetend.

Pure and mixed circulating decimals are generally written

down only to the end of the first period, a dot being placed over the first and last figures of that period.

Thus $\cdot\dot{3}$ represents the pure circulat⁸. dec¹. $\quad\cdot 333..$
$\cdot\dot{3}\dot{6}$ .... $\cdot 3636..$
$\cdot\dot{6}3\dot{9}$ .... $\cdot 639639..$
$\cdot 13\dot{8}$ ....mixed .... $\cdot 1388.$
$\cdot 013\dot{6}$ .... $\cdot 0113636..$

100. *Pure Circulating Decimals may be converted into their equivalent Vulgar Fractions by the following Rule.*

RULE. Make the period or repetend the numerator of the fraction, and for the denominator put down as many *nines* as there are figures in the period or repetend.

This fraction, reduced to its lowest terms, will be the fraction required in its simplest form.

Ex¹. Reduce the following pure circulat⁸. dec¹⁸., $\cdot\dot{3}, \cdot\dot{2}\dot{7}$, $\cdot\dot{8}5714\dot{2}$, to their respective equivalent vulgar fractions.

By the Rule, $\dot{3} = \dfrac{3}{9} = \dfrac{1}{3}; \quad \dot{2}\dot{7} = \dfrac{27}{99} = \dfrac{3}{11}.$

$\dot{8}5714\dot{2} = \dfrac{857142}{999999} = \dfrac{142857 \times 6}{142857 \times 7} = \dfrac{6}{7}.$

101. *Mixed Circulating Decimals may be converted into their equivalent Vulgar Fractions by the following Rule.*

RULE. Subtract the figures which do not circulate from the figures taken to the end of the first period, as if both were whole numbers.

Make the result the num⁽·⁾; and write down as many *nines* as there are figures in the circulating part, followed by as many *zeros* as there are figures in the non-circulating part, for the denominator.

Ex⁸. Reduce the following mixed circulat⁸. dec¹⁸., $\cdot 1\dot{4}$, $\cdot 013\dot{8}, \cdot 241\dot{8}$, to their respective equivalent vulgar fractions.

By the Rule, $\cdot 1\dot{4} = \dfrac{14-1}{90} = \dfrac{13}{90}; \quad \cdot 013\dot{8} = \dfrac{138-13}{9000} = \dfrac{125}{9000}$

$= \dfrac{1}{72}; \quad \cdot 241\dot{8} = \dfrac{2418-2}{9990} = \dfrac{2416}{9990} = \dfrac{1208}{4995}.$

102. In the Addition and Subtraction of circulating decimals, where the result is only required to be true to a certain number of decimal places, it will be sufficient to carry on the circulating part to two or three decimal places more than the number required; taking care that the last figure

ARITHMETIC.

retained be increased by 1, if the succeeding figure be 5, or greater than 5. In the Multiplication and Division, however, of circulating decimals, it is always preferable to reduce the circulating decimals to Vulgar Fractions, and having found the product or quotient as a Vulgar Fraction, then, if necessary, to reduce the result to a decimal.

Ex. LXI.

Reduce to circulating decimals: (1) $\frac{2}{3}$; $\frac{1}{9}$; $\frac{6}{7}$; $\frac{7}{12}$; $\frac{11}{15}$, (2) $6\frac{3}{81}$; $7\frac{5}{37}$; $100\frac{1}{11}$; $2\frac{1}{7}$. (3) $11\frac{57}{105}$; $23\frac{17}{1375}$.

Reduce to their equivalent vulgar fractions: (4) $\cdot\dot{2}$; $\cdot\dot{0}\dot{5}$; $\cdot\dot{1}\dot{8}$; $\cdot\dot{1}5\dot{6}$; $\cdot0\dot{2}70\dot{2}\dot{7}$; $\cdot\dot{2}8571\dot{4}$. (5) $\cdot 5\dot{6}\dot{6}$; $\cdot 7\dot{4}\dot{3}$; $\cdot 20\dot{4}3\dot{5}$; $19\cdot 30\dot{5}$; $20\cdot 0291\dot{6}$. (6) $6\cdot 1\dot{8}1531\dot{5}\dot{3}$; $15\cdot\dot{6}9230\dot{7}\dot{6}9230\dot{7}$.

Find the value correct to six places of dec^{ls}. of (7) $4\cdot\dot{3} + 16\cdot\dot{4}\dot{5} + 75\cdot 735\dot{2}$. (8) $3\cdot 2\dot{3} + 26\cdot 79\dot{6} + 7\cdot\dot{4}1\dot{3}$. (9) $3\cdot 856\dot{4} - 2\cdot 038\dot{7}$. (10) $52\cdot 8\dot{6} - 8\cdot 37235$.

Find the value of (11) $7\cdot\dot{6} \times 5\cdot\dot{3}$; $\cdot 35\dot{1} \times \cdot 73\dot{6}$; $\cdot 1\dot{3} \times \cdot 2 \times \cdot\dot{4}$. (12) $6\cdot\dot{7} \div 2\cdot\dot{6}$; $\cdot 262\dot{7} \div 1\cdot 92\dot{6}$; $\cdot 37\dot{1} \div 5$; $42\cdot 0463 \div 1\cdot 3\dot{6}$.

REDUCTION OF DECIMALS.

103. *To reduce a decimal of any denomination to its proper value.*

RULE. Multiply the given decimal by the number of units of the next lower denomination which make one of the given denomination, and point off for decimals as many figures in the product, beginning from the right hand, as there are figures in the given decimal.

The figures on the left of the decimal point will represent the whole numbers in the next denomination.

Proceed in the same way with the decimal part for that denomination, and so on.

Ex. 1. Find the value of $\cdot 4625$ of £1.

By the Rule,

£.
$\cdot 4625$
20
———
$9\cdot 2500 s.$
12
———
$3\cdot 0000 d.$

$£\cdot 4625 = 9s.$ **3d.**

By fractions,

$£\cdot 4625 = \left(\dfrac{4625}{10000} \times 20\right)s. = \left(\dfrac{92500}{10000}\right)s.$

$= 9\frac{25}{100}s. = 9s. + \left(\dfrac{25 \times 12}{100}\right)d.$

$= 9s. + \dfrac{300}{100}d. = 9s,\ 3d.$

DECIMALS.

Note. If the **quantity**, the **value** of whose decimal part is to be found, be **a compound** quantity, it must be reduced to *one* denomination before the rule is applied.

Ex. 2. Find the value of 7·405 of 15 mi., 5 fur., 31 po.

```
                           po.
   15 mi., 5 fur., 31 po. = 5031
                           7·405
                          ───────
                           25155
                          20124
                          35217
                         ────────
                         37254·555 po
                               5½
                            ───────
                            2·775
                             ·2775
                            ───────
                            3·0525 yds.
                               36
                            ───────               po.  yds.  in.
                            3150        ∴ val⁰ = 37254   3   1·89
                            1575                  mi. fur. po. yds. in.
                           ───────           or = 116   3   14   3   1·89
                           1·8900 in.
```

Ex. 3. Find the value of ·5416 of 4½ cwt.

1st method.

cwt. lbs.
4½ = 450   ·541666
              450
         ─────────   2d. method.
          270833 00
         21 6666 4   ·5416 of 4½ = ((5416 − 541)/9000 of 9/2) cwt.
         ─────────
         243·749700 lbs.   = (4875/9000 × 9/2) cwt. = (13/24 × 9/2 × 100) lbs.
                  16
         ─────────
         11·995200 oz.   = 243 lbs., 12 oz.

∴ value = 243 lbs., 12 oz. nearly. The 2nd method is the better one in most cases.

Ex. LXII.

Find the value of,

(1) ·75 of $1.   (2) ·875 of $5.   (3) ·625 of $1.
(4) ·625 of 1 cwt.   (5) ·375 of a mi.   (6) ·175 of a ton.
(7) ·46875 of £2. 10s.   (8) ·0625 of 7s. 6d.
(9) 3·175 of 1 lb. Troy.   (10) 4·65 of 4½ ac.
(11) 10·04 of 2½ ro.   (12) 2·56 of 10s. 11¼d.
(13) 5·00875 of 3 wks. 4 dys.   (14) 16·504 days.

(15) $3{\cdot}05$ of 5 lbs. $2\dot{9}$).   (16) $3{\cdot}00\dot{8}\dot{5}$ of £4. 1s.
(17) $7{\cdot}03\dot{4}$ of 1 ac., 3 ro., 5 po.
(18) $5{\cdot}00\dot{5}$ of 16 lbs., 1 oz., 6 grs. Troy.
(19) $\cdot\dot{3}$ of $2. (20) $\cdot 5\dot{4}$ of 16s. 6d. (21) $\cdot 2\dot{4}\dot{3}$ of a ton
(22) $6{\cdot}8\dot{3}$ of £5. (23) $2{\cdot}38\dot{3}$ of $2\frac{1}{2}$ lbs. T'y. (24) $6{\cdot}\dot{2}$ of ac. yd.
(25) $18{\cdot}\dot{7}\dot{2}$ of an ac. (26) $2{\cdot}06\dot{3}$ of 1000 guineas.
(27) £$\cdot 634375 + \cdot 025$ of 25s. + $3{\cdot}1\dot{6}$ of 30s.
(28) $\cdot \dot{6}$ of an ac. + $\cdot 625$ of a ro. — $\frac{1}{11}$ po.
(29) $6{\cdot}\dot{7}1428\dot{5}$ of 1s. 9d. — $\cdot 08\dot{3}\dot{3}$ of £7. 4s. + $\cdot 251190476$ of 6s. 8d.

104. *To reduce* **a number** or *fraction of one or more denominations to the* **decimal** *of another denomination of the same kind.*

Rule. Reduce the given **number** or fraction to a fraction of the proposed denomination; and then reduce this fraction to its equivalent decimal.

Ex. 1. Reduce $\frac{2}{5}$ of £1 to the decimal of a guinea.
$\frac{2}{5}$ of £1 = $\frac{40}{5}$s. = 8s. 1 guin^a. = 21s., ∴ fraction req^d. = $\frac{8}{21}$.
Now $8 \div 21 = \cdot 380952\dot{3}\dot{8}\ldots$, ∴ dec^l. req$_d$. = $\cdot \dot{3}8095\dot{2}$.

Ex. 2. What decimal of £2 is 11s. $9\frac{3}{4}d.$?
$11s.\ 9\frac{3}{4}d. = 567q.;\ £2 = 1920q.$
∴ fract^n. req^d. = $\frac{567}{1920} = \frac{189}{640}$, ∴ dec^l. req^d. = $189 \div 640 = \cdot 2953152$;
or thus,

```
4 | 3·00
  |_____        We first reduce $\frac{3}{4}d.$ to the dec$^l$. of 1$d.$, by div$^s$.
12| 9·75        3$d.$ by 4, which = ·75$d.$, next 9·75$d.$ to the dec$^l$.
  |_____        of 1$s.$, by div$^s$. by 12, which = ·8125$s.$, then
4,0|11·8125     11·8125$s.$ to the dec$^l$. of £2, by div$^s$. by 40
   |_____       which = £·2953125.
   |·2953125
```

Ex. LXIII.

Reduce,
(1) 1 qr., 5 lbs. to the dec^l. of a **cwt.**
(2) $2.50 to the dec^l. of $10.
(3) 3 hrs., 30' to the dec^l. of a day.
(4) 3 ro., 11 per. to the dec^l. of an acre.
(5) $6\frac{1}{4}d.$ to the dec^l. of a shilling.
(6) $3\frac{1}{2}$ in. to the dec^l. of 2 furlongs.
(7) 2 oz., 13 dwt. to the dec^l. of a lb.
(8) 4 lbs., 2 so. to the dec^l. of an oz.

(9) 2 sq. ft. 73 in. to the dec¹. of a sq. yd.
(10) 1 lb. Troy to the dec¹. of a lb. Avoir.
(11) 10s. 9d. to the dec¹. of a £.
(12) 17s. 7d. to the dec¹. of a £.
(13) 2 wks., 6¼ dys. to the dec¹. of 4 dys. 3 hrs.
(14) 2 lbs., 14 oz. to the dec¹. of 18 lbs.

Ex. LXIV.
MISCELLANEOUS EXAMPLES.
PAPER I.

(1) Define a unit; a number. Into what classes are numbers divided? Explain the difference between them. Define Notation and Numeration.

(2) Write down in words the following numbers:
70340;   125004321;   5607605213403;
and express by numbers eight hundred and ten thousand four hundred and one; sixty-four billions two millions six hundred and forty-six thousand and two.

(3) (1) Add together one million eighteen thousand two hundred and sixty-nine; twenty thousand nine hundred and seventy-nine; one hundred millions one thousand and fifty; fifty-four billions three thousand; four hundred millions and six; nine hundred and ninety-nine thousand nine hundred and ninety. (2) Subtract 300725 from 400001.

Explain clearly why you carry 1 when you borrow ten.

(4) (1) Multiply 268936785 by 5689, and verify by division. (2) Divide 27027027027 by 6974, and verify by multiplication.

(5) The product is 99626417315464, the multiplier 72568; what is the multiplicand?

(6) In 12 mi., 2 fur., 6 per., how many inches? Shew that your result is correct.

PAPER II.

(1) When is a number said to be a *multiple* of another number? What is a *common multiple*? What is the *least common multiple* of two or more numbers? Find L. C. M. of 27, 36, 42, 48.

(2) Explain the meaning of the signs, $+$, $-$, $=$. When can questions in *Addition* be performed by *Multiplication*.

(3) A cask is required to be **exactly filled** by any one of the following measures: 1 pint, 2 pints, 3 pints, 5 pints, 6 pints, or 9 pints; find the **smallest cask** for the purpose.

(4) The forewheel of a waggon is 8 feet round, and the hind-wheel fourteen; how many feet will the waggon travel over before each wheel shall have made a number of complete turns? How often will this happen in 1000 feet?

(5) The length and cost of building the undernamed Canadian Canals, were as follows: The Rideau Canal, $126\tfrac{1}{4}$ miles, \$4380000; the St. Lawrence Canal, $40\tfrac{1}{2}$ miles, \$8550000; the Ottawa Canal, $10\tfrac{1}{2}$ miles, \$1500000; the Chambly and St. Ours Lock, $11\tfrac{1}{2}$ miles, \$550000; the Welland Canal and feeder, $50\tfrac{1}{2}$ miles, \$7000000; the Burlington and Desjardins Bridge cost \$560000. Find (1) the total length of the above canals, (2) their total cost, and (3) the average cost per mile, excluding the Burlington and Desjardins Bridge.

(6) Define a vulgar fraction? Distinguish between a vulgar and decimal fraction? Give an example of the different kinds of vulgar fractions?

PAPER III.

(1) Simplify (1) $2\tfrac{1}{4} (\tfrac{1}{8} + \tfrac{3}{8}) + \tfrac{1}{8} (\tfrac{3}{8} - \tfrac{1}{2})$.

(2) $2\tfrac{1}{4} \{(\tfrac{1}{8} + \tfrac{3}{8}) + \tfrac{1}{8} (\tfrac{3}{8} - \tfrac{1}{2})\}$.

(2) A person who owns $\tfrac{1}{2}$ of a steam-vessel, sells $\tfrac{2}{3}$ of his share for \$15000; what is the remaining part of his share worth?

(3) Simplify (1) $\dfrac{7}{12} (8\tfrac{1}{4} - 2\tfrac{1}{2}) - \tfrac{1}{2} (\tfrac{3}{8} - \tfrac{1}{16})$.

(2) $\dfrac{7}{12} \{(8\tfrac{1}{4} - 2\tfrac{1}{2}) - \tfrac{1}{2} (\tfrac{3}{8} - \tfrac{1}{16})\}$

(4) A clerk copied ·55 of £5 instead of 5·5 of £5, what was the amount of the error?

(5) It takes 87 yds. of carpet, 1·25 yd. wide, to cover a room, how many more yds. will it take, if the width be ·75 yd?

(6) A gave ·5 of an orange to B; ·3 of what remained to C: how much of the orange had A left for himself?

PAPER IV.

(1) A drover sold $\tfrac{1}{2}$ of his flock to A, $\tfrac{1}{2}$ of the remainder to B, and the rest to C. How many had he at first, supposing C got 32?

(2) Add together $13\tfrac{1}{2}$, $56\tfrac{3}{4}$, and $14\tfrac{4}{5}$ by vulgar and decimal fractions, and shew that the results coincide.

(3) The product of two decimals is ·033372; one of them is 2·7; find the other.

(4) Add together £27. 6s. 9½d., $17.22, £19. 5s. 8d., $198.05, £3. 12s. 7d. The answer to be in dec¹. currency.

(5) At a football match there were ·875 as many on one side as on the other, and the players on both sides were equal in number to ·625 of the lookers on; if there were 21 on the smaller side, how many were playing on the other side, and how many were looking on?

(6) If in a cricket match one side scores ·014 of 1⅔ of $\dfrac{1\frac{1}{5}}{5\frac{2}{3}}$ of ⅘ of 4·5 of $\dfrac{\frac{3}{4}+\frac{1}{2}}{\frac{3}{4}-\frac{1}{2}}$ of 71¾ of the score made by the other side; which side wins?

PAPER V.

(1) C owes B ·6 of what B owes A, B gives C 5s to put the accounts between them all straight. What is B's debt to A?

(2) Out of a bag of silver, I take 50s. more than ·5 of the whole sum which it contained; then 30s. more than ·2 of what then remained; and then 20s. more than ·25 of what then remained; after this 10s. remained. What did the bag contain at first?

(3) A bath, containing 286 cub. yds. has two inlets A and B, which respectively supply 26 cub. yds. in 3¼ hrs., and 12½ cub. yds. in 2½ hrs.; and also an outlet C, which discharges 11·375 cub. yds. in 1¾ hrs., if the bath be empty, and A and C open for 12 hrs., and then B also open, in what time will ·75 of it be filled?

Make out the following bills:

(4) 500 envelopes at 44 cents per 100, 3 boxes of elastic bands at 33 cents per box, ½ a gross of penholders at 19 cents per doz. 2½ reams of foolscap at 21 cents per quire, 4 dozen quill pens at 9 cents per doz., 13 note books at 27 cents each, and 250 official envelopes at 48 cents per 100.

(5) A loin of lamb (7½ lbs.) at 10 cents per lb., a haunch of mutton (19½ lbs.) at 8 cents per lb., a pork ham (18 lbs.) at 15 cents per lb., 5½ lbs. of suet at 10 cents per lb., and 9 chops at 4 cents each.

(6) 17 yds. calico at 19 cents per yd., 25½ yds. at 55 cents per yd., 34½ yards of flannel at 60 cents per yd., 14 pairs of stockings at 38 cents a pair, and 5 pairs of gloves at $12 per doz.

SECTION V.

RATIO AND PROPORTION.

105. Numbers are divided into two classes, ABSTRACT and CONCRETE. One, or the number one, when the unit does not refer to any particular object, is an *abstract* number. One, in the expression one pound, when the unit refers to a particular object, viz. "a pound," is a *concrete* number.

106. We may ascertain the relation which one abstract number bears to another abstract number or one concrete number to another concrete number of the *same kind*, by expressing the first number as a fraction of the second: thus the relations which 12 bears to 3, and 3 to 12 are expressed by the fractions $\frac{12}{3}$ or $\frac{4}{1}$, and $\frac{3}{12}$ or $\frac{1}{4}$; also the relations which 12s. bears to 3d., or 144d. to 3d., and 3d. to 12s., or 3d. to 144d. are expressed by the fractions $\frac{144}{3}$ or $\frac{48}{1}$, and $\frac{3}{144}$ or $\frac{1}{48}$.

107. The relation of one number to another in respect of magnitude is called RATIO. The Ratio of one number to another may be expressed by the fraction which the first is of the second.

108. The Ratio of one number to another is often denoted by placing a colon between them. Thus the ratios of 12 to 3 and 3 to 12 are denoted by 12 : 3 and 3 : 12. Hence it follows that $12 : 3 = \frac{12}{3}$, and $3 : 12 = \frac{3}{12}$.

109. The two numbers which form a Ratio are called its *terms;* the former term is called the ANTECEDENT, the latter the CONSEQUENT. Since 3d. reduced to the fraction of 12s. $=\frac{3}{144}$, it is clear that when we have two concrete numbers of the same kind, but of different denominations, we must, in order to find their ratio, **reduce** them to one and the same denomination, and may then **treat** them as abstract numbers

RATIO AND PROPORTION. 111

110. When **two Ratios are equal**, in other words, when they can be expressed by the same fraction, they are said to form a PROPORTION, and the four numbers are called PROPORTIONALS. Thus the ratio of 8 to 9 is equal to that of 24 to 27, for $8:9 = \frac{8}{9}$, and $24:27 = \frac{24}{27} = \frac{8}{9}$. The Ratios being equal, Proportion exists among the numbers 8, 9, 24, 27; and those numbers are Proportionals.

111. The existence of Proportion between the numbers 8, 9, 24, 27 is denoted thus, $8:9 = 24:27$, or $8:9::24:27$, which is usually read thus, 8 is to 9 as 24 is to 27.

112. In any Proportion, as $8:9::24:27$, *the product of the* 1st *and* 4th, *i.e. the extreme terms* = *the product of the* 2nd *and* 3rd, *i.e. the mean terms;*

$$\frac{8}{9} = \frac{24}{27}; \therefore \frac{8}{9} \times 9 \times 27 = \frac{24}{27} \times 9 \times 27, \text{ or } 8 \times 27 = 24 \times 9.$$

113. *If four numbers be proportionals when taken in a certain order, they will also be proportionals when taken in the contrary order.* For instance, 8, 9, 24, 27 are proportionals;

$$\therefore \frac{8}{9} = \frac{24}{27}; \therefore 1 \div \frac{8}{9} = 1 \div \frac{24}{27}; \text{ or } \frac{9}{8} = \frac{27}{24}, \text{ or } \frac{27}{24} = \frac{9}{8};$$

$$\therefore 27:24::9:8.$$

114. *If any three terms of a proportion be given, the remaining term may always be found.*

For since in any Proportion

1st term × 4th term = 2nd term × 3rd term;

$$\therefore \text{1st term} = \frac{\text{2nd} \times \text{3rd}}{\text{4th}}, \text{ 2nd term} = \frac{\text{1st} \times \text{4th}}{\text{3rd}},$$

$$\text{3rd term} = \frac{\text{1st} \times \text{4th}}{\text{2nd}}, \text{ 4th term} = \frac{\text{2nd} \times \text{3rd}}{\text{1st}}.$$

Ex. 1. Find **the 4th term in the** proportion 2, 3, 18.

$\mathbf{2:3::18}:$ 4th term$; \therefore$ 4th term $= \frac{3 \times 18}{2} = 27.$

Ex. 2. Find **the 2nd term** in the proportion 8, 32, and 24.

$\mathbf{8}:$ 2nd term $::32:24; \therefore$ 2nd term $= \frac{8 \times 24}{32} = 6.$

ARITHMETIC.

Ex. LXV.

Find the 4th term in each of the following proportions:

(1) $4 : 9 :: 12 :$ (2) $32 : 9 :: 24 :$
(3) $4 : 6 :: 10 :$ (4) $\frac{1}{2} : \frac{1}{3} :: \frac{1}{4} :$
(5) $\cdot 05 : \cdot 8 :: \cdot 79 :$ (6) $3 : 10 :: 4\cdot 5 :$

Find the 2nd term in each of the proportions:

(7) $\frac{5}{7} : :: \frac{19}{21} : \frac{19}{27}.$ (8) $1\cdot 2 : :: 1\cdot 3 : \cdot 39.$

Find the 1st term in each of the proportions:

(9) $: \frac{9}{14} : \frac{7}{15} : \frac{1}{3}.$ (10) $: 4\cdot 2\dot{2} : 17\cdot 6 : 23\frac{7}{9}.$

RULE OF THREE.

115. The RULE OF THREE is a method by which we are enabled, from three numbers which are given, to find a fourth which shall bear the same ratio to the third as the second to the first; in other words, it is a Rule by which, when three terms of a proportion are given, we can determine the fourth.

116. RULE. Find out of the three quantities which are given, that which is of the same kind as the fourth or required quantity; or that which is distinguished from the other terms by the nature of the question: place this quantity as the third term of the proportion.

Now consider whether, from the nature of the question, the fourth term will be greater or less than the third; if greater, then put the larger of the other two quantities in the second term, and the smaller in the first term; but if less, put the larger in the first term and the smaller in the second term.

Take care to reduce the first and second terms to one and the same denomination, and also to reduce the third so that it may be wholly in one denomination; remembering, however, that if the quantities involved be all of the same kind, it is unnecessary to reduce all the three terms to the same denomination, but only the first and second terms to one and the same denomination, and the third to a single denomination, which will not necessarily be the same as the former. When the terms have been properly reduced, multiply the second and third together, and divide by the first, treating all three as abstract numbers. The quotient will be the answer to the question, in the denomination to which the third term was reduced.

RULE OF THREE. 113

If 19 bushels of potatoes cost $15.20, how many bushels can be bought for $83.20? Since 19 bush. is of the same kind as the req^d. term, viz., bus., we make 19 bus. the 3rd. term; since $83.20 can buy more bus. than $15.20, we make $83.20 the 2nd. term, and $15.20 the 1st. term:

$$\$ \text{ c.} \quad \$ \text{ c.} \quad \text{bus.}$$
$$15.20 : 83.20 :: 19 : \text{no. of bus. req}^d.$$
or 1520 cts. : 8320 cts. :: 19 bus. : of bus. req^d.

$$\therefore \text{ no. of bus. req}^d. = \frac{8320 \times 19}{1520} = 104.$$

Ex. 2. A gentleman hired a servant for the year 1865 for £32. 13s. 11½d., the man left his service on the evening of the last day of June: what amount of wages ought to be paid to him?

From Jan. 1 to June 30, both included, there are (31 + 28 + 31 + 30 + 31 + 30) days = 181 days;

We place £32. 13s. 11½d., the given quantity of the req^d. kind, in the 3rd. term; wages for 181 days will be *less* than wages for 365 days, $\therefore$ place 181 days in the 2nd. term, and 365 days in the 1st. term.

$$\text{days.} \quad \text{days.} \quad £ \quad s. \quad d.$$
$$\therefore 365 : 181 :: 32 \ 13 \ 11\tfrac{1}{2} : \text{req}^d. \text{ am}^t. \text{ of wages.}$$
or 365 days : 181 days :: $31390q$ . . : . . . . in q.

$$\therefore \text{req}^d. \text{ am}^t. \text{ of wages} = \frac{31390 \times 181}{365} q. = £16. \ 4s. \ 3\tfrac{1}{2}d.$$

Ex. 3. A bankrupt can pay 9s. 0½d. in the £, and his assets amount to £1069. 3s. 6½d.; find the amount of his debts.

For every asset of 9s. 0½d. he owes £1, $\therefore$ place £1 in the 3rd term.

$$9s. \ 0\tfrac{1}{2}d. : £1069. \ 3s. \ 6\tfrac{1}{2}d. :: £1 : \text{am}^t. \text{ of debts in £'s,}$$
or 217 half-pence : 513205 half-pence :: £1 : am^t. of debts in £'s;

$$\therefore \text{am}^t. \text{ of debts in £'s} = \frac{513205}{217} = 2365.$$

Ex. 4. If ·0625 of 1 lb. cost ·458s.; what will ·075 of a ton cost?

$$\text{lb.} \quad \text{ton} \quad s.$$
$$·0625 : ·075 :: ·458 : \text{req}^d. \text{ price in shillings,}$$
$$\text{lb.} \quad \text{lbs.} \quad s.$$
or $·0625 : ·075 \times 20 \times 112 :: ·458 : \text{req}^d.$ price in shillings

$$\therefore \text{price} = \frac{·458 \times ·075 \times 20 \times 112}{·0625} s. = £61. \ 11s. \ 1·248d.$$

ARITHMETIC.

Ex. 5. A owned $\frac{8}{11}$ths of a ship, and sold $\frac{5}{11}$ of $\frac{2}{3}$ of his share for £12$\frac{4}{33}$; what was the value of $\frac{1\frac{1}{4}}{4\frac{1}{4}}$ of $\frac{2}{5}$ths of the vessel?

$\frac{5}{11}$ of $\frac{2}{3}$ of $\frac{4}{17}$: $\frac{1\frac{1}{4}}{4\frac{1}{4}}$ of $\frac{2}{5}$ :: £12$\frac{4}{33}$: req^d. value in £'s,

or $\dfrac{2 \times 4}{11 \times 3 \times 17} : \dfrac{5}{4} \times \dfrac{4}{17} \times \dfrac{2}{5}$ :: £$\dfrac{400}{33}$: req^d. value in £'s;

∴ req^d. value in £'s $= \dfrac{400 \times 2}{33 \times 17} \times \dfrac{11 \times 3 \times 17}{2 \times 4} = 100.$

Note 1. There are certain examples in which, at first sight, more than three terms appear to be given, but they, in certain cases, come under this Rule, as in the following instances:

Ex. 6. If the carriage of 5 cwt., 7 lbs., for 84 miles cost £3. 18s. 4d., what will it cost to have 21 cwt., 1 qr., 14 lbs. carried the same distance?

84 miles may be left out of consideration, the distance in both cases being the same.

∴ 5 cwt., 7 lbs. : 21 cwt., 1 qr., 14 lbs. :: £3. 18s. 4d. : req^d. cost; whence, req^d. cost = £16. 10s. 8$\frac{3}{4}$d. $\frac{8}{9}$q.

Ex. 7. If 12 men can reap a field in 4 days, in what time can the same work be performed by 32 men?

32 men require less than 4 days to perform the work ·
∴ 32 : 12 :: 4 days : req^d. time in days;

∴ req^d. time $= \dfrac{12 \times 4}{32}$ days $= 1\frac{1}{2}$ days.

Note 2. Examples such as the following are easily worked by Rule of Three.

Ex. 8. A gentleman after paying an income-tax of 7d. in the £, has £248. 10s. 8d.; what was his gross annual income?
After paying inc^e. tax on £1, he had £1 less 7d., or 19s. 5d.

∴ 19s. 5d. : £248. 10s. 8d. :: £1. : req^d. income; whence, req^d. income = £256.

Ex. 9. A hare, pursued by a greyhound, was 130 yards before him at starting; whilst the hare ran 5 yards the dog ran 7 yards: how far had the hare gone when she was caught by the greyhound?

Since the dog gains 2 yds. on every 5 yds. which the hare

RULE OF THREE.

runs, we require to find how many yards the hare must run for the dog to gain 130 yds.

∴ 2 yds. : 130 yds. :: 5 yds. : no. of yds. the hare must run ;

$$\therefore \text{no. of yds. req}^d = \frac{130 \times 5}{2} = 325.$$

Ex. LXVI.

(1) If 8 bushels of wheat cost $16, what will 24 bushels cost at the same rate?

(2) If 2 bushels of oats cost $1.10, how much will 33 bushels cost?

(3) If 9 bushels clover seed cost $36, how much will 4 bus. 20 lbs. cost?

(4) When oats are selling at 55 cents a bushel; how many bushels can be bought for $21.25?

(5) The price of a bushel of pease being 84 cents; how many bushels can be bought for $17.20?

(6) Find the value of a silver salver, weighing 21 lbs., 4 oz. at 6s. 5d. an oz.

(7) How much cheese at 16 cts. per lb. can be bought for $462.36?

(8) A bankrupt, who owes $23856, can pay $10496.64; what will be the dividend in the $?

(9) A pensioner received $106.14 for the year 1864; find the amount of his daily pension.

(10) 1 mile of road cost $393.75; what will 20 mi., 5 fur., 22 yds. of the same kind of road cost?

(11) What weight of sugar may be bought for $449.23, when the cost of 6 cwt., 2 qrs. is $133.12.

(12) The taxes on a house rated at $183.75 amount to $32.15; the taxes on another house in the same village amount to $286.66½; find the rateable value of the 2nd house.

(13) A bankrupt's debts amount to $10000, and his property to $3875, what will each of his creditors lose in the $?

(14) A ship was provisioned for a crew of 84 men for 5 months; how much longer would the provisions last, if a crew of only 60 men were taken on board?

(15) A merchant exchanged 1134 yds. of velvet for 5313

ARITHMETIC.

yds. of silk at 3s. 4½d. a yd.; find the value of the velvet a yd.?

(16) What are the effects of a bankrupt worth, whose debts amount to £3057. 12s., and who can pay 17s. 6d. in the £?

(17) A man on the average walks over 10 ft., 8 in. in 4 steps, what number of steps will he take between two places, a distance of 1 mi., 1280 yds. apart?

(18) If 31 ac., 3 ro., 9 po., 21 yds. of ground cost £3025 12s. 4½d., what will be the price of 49 ac., 3 ro., 38 po., 2¾ yds. of ground of the same quality?

(19) A bankrupt pays 59 cts. in the $; what will be lost on a debt of $13675?

(20) How many minutes must a boy, who runs 6 mi. an hour, start before another boy, who runs 7½ mi. an hour in order that they may be together at the end of 10 mi.?

(21) Two boats start in a race, and one of them gains 5 ft. upon the other in every 55 yds.; how much will it have gained at the end of half a mile?

(22) How many pairs of mits at 45 cts. a pair should be exchanged for 36 dozen pairs of stockings at 55 cts. a pair?

(23) How many men would perform in 168 days a piece of work, which 108 men can do in 266 days?

(24) If an incorporated village be rated at $12571.87½ and a rate be granted of $419.06¼; how much is the rate in the $? How much will be paid by a house rated at $1734.37½.

(25) A gentleman's income in 1863 was $2500, out of which he saved $994.37½; find his average daily expenditure.

(26) If 100 men can finish a piece of work in 27 days, how many men will finish it in 20 days?

(27) A special train on the Grand Trunk Railway, which travels at the uniform rate of 44 ft. in a second, leaves Belleville for Toronto, a distance of 109 miles, at 8 o'clock A. M.; at what time will the train reach Toronto.

(28) A bankrupt owes to one creditor a certain sum, to each of two others $1250, to each of three others $810; his property is worth $1718.75, and he can pay 22 cts. in the $. How much will the first creditor lose?

(29) If, when wheat is 42s. a qr. (8 bus.), the 4 lb.

RULE OF THREE.

costs 4½d., what ought the 4 lb. loaf to cost when wheat is 70s. a qr. ?

(30) In what time ought 10 men to perform the same work, which 5 men and 5 boys can perform in 15 days, it being given that 3 men can perform the same amount of work as 5 boys ?

(31) Find a 4th proportional to 1 lb., 10 oz., 10 dwts. ; 1 oz. ; and £6. 3s. 9d.

(32) How much might a person have spent in Jan., 1864, who wished to save in that year $250 out of an income of $2034.50 ?

(33) A person, after paying an income-tax of 6d. in the £, has £877. 10s. left, find his original income.

(34) Find (1) the income which pays £29. 3s. 4d. tax at the rate of 7d. in the £ ; (2) the income from which, after paying tax at the same rate, the remainder is £932.

(35) A piece of gold at £3. 17s. 10½d. per oz. is worth £150 ; what will be the worth of a piece of silver of equal weight at 54s. 6d. per lb. ?

(36) A certain piece of work was to be done by 25 men in 16 days; after 4 days 15 men go away ; how long will it take the rest of the men to finish the work ?

(37) A person after paying for the 1st half of a year an income-tax of 1 ct. in the $, and for the 2nd half one of 1½ cts. in the $ on his income, has $1855 left ; what was the income on which he paid ?

(38) If $\frac{4}{5}$ of a qr. of wheat cost 54s., what will be the price of $\frac{4}{5}$ of a bus. ?

(39) If $1\frac{2}{3}$ of a cwt. cost £7 3s., what will $\frac{8}{11}$ of a ton cost ?

(40) If $\frac{1}{100}$ of $\frac{3}{4}$ of $2\frac{1}{2}$ of 40 lbs. of beef cost $1\frac{3}{5}d.$, how many lbs. can be bought for £1. 6s. 6d. ?

(41) A clock marks the true time on Sunday morning at 6 o'clock, and on Tuesday at noon it has gained 24 minutes, what will be the true time when it shews 1 o'clock on Saturday afternoon ?

(42) The hour and minute hands of a watch are together at 12 o'clock, when will they next be together ?

(43) If 5 lbs. of sugar cost ·0703125 of $4, what will ·0625 cwt. of the same sugar cost ?

(44) A certain piece of work can be done in 18 days by

118   ARITHMETIC.

4 men, 7 women, or 9 boys; how long will the same work occupy 5 men, 4 women, and 2 boys?

(45) If after selling ⅜ths of an estate, I sell ½ of ⅞ of the remainder for 1½ of ⅔ of £600⅔, what is the value of ⅔rds of it?

(46) What will be the value of a gold cup weighing 2·683 lbs.; when 1 oz. of it is worth £4·09?

(47) 4 men and 5 boys earn $22.12 in 7 days, and 3 men and 8 boys earn $28.98 in 9 days; in what time will 12 men and 12 boys earn $186.48?

(48) A can do a piece of work in 5 hours, B in 9 hours, and C in 15 hours. How long will it take C to finish the work, after A has worked at it for 40 minutes, and B for 1½ hours?

(49) If a garrison of 1500 men have provisions for 13 mo., how long will their provisions last, if at the end of 2 mo. they be reinforced by 700 men?

(50) Two men start at 8.30 A.M., one from Toronto and the other from Whitby, a distance of 30 miles, and they approach each other at the rates of 4½ and 3 miles an hour; at what time will they meet, and at what distance from a place, which is 2 miles nearer to Toronto than Whitby?

(51) Two trains respectively 210 feet and 180 feet in length are going in opposite directions, the first at the rate of 24 miles per hour, and the other at the rate of 27 miles per hour; find how long they will take to pass each other?

DOUBLE RULE OF THREE.

117. The DOUBLE RULE OF THREE is a shorter method of working out such questions as would require two or more applications of the Rule of Three.

118. For the sake of convenience, we may divide each question in the Double Rule of Three into two parts, the *supposition* and the *demand*; the supposition being the part which expresses the conditions of the question, and the demand the part which mentions the thing demanded or sought. In the question, "If the carriage of 15 cwt. for 17 miles cost $21, what would the carriage of 21 cwt. for 16 miles cost?" the words "if the carriage of 15 cwt for 17 miles cost $21," form the supposition; and the words "what would the carriage of 21 cwt. for 16 miles cost?" form the demand.

DOUBLE RULE OF THREE.

Adopting this distinction, we may give the following Rule for working out examples in the Double Rule of Three.

119. RULE. Take from the supposition that quantity which corresponds to the quantity sought in the demand; and write it down as a third term. Then take one of the other quantities in the supposition and the corresponding quantity in the demand, and consider them with reference to the third term *only* (regarding each other quantity in the supposition and its corresponding quantity in the demand as being equal to each other); when the two quantities are so considered, if from the nature of the case, the fourth term would be greater than the third, then, as in the Rule of Three, put the larger of the two quantities in the second term, and the smaller in the first term; but if less, put the smaller in the second term, and the larger in the first term.

Again, take another of the quantities given in the supposition, and the corresponding quantity in the demand; and retaining the same third term, proceed in the same way to make one of those quantities a first term and the other a second term.

If there be other quantities in the supposition and demand, proceed in like manner with them.

In each of these statings reduce the first and second terms to the same denomination. Let the common third term be also reduced to a single denomination if it be not already in that state. The terms may then be treated as abstract numbers.

Multiply all the first terms together for a final first term, and all the second terms together for a final second term, and retain the former third term. In this final stating multiply the second and third terms together and divide the product by the first. The quotient will be the answer to the question in the denomination to which the third term was reduced.

Ex. 1. If 5 men earn £18. 15s. in 12 weeks, how much will 16 men earn in 20 weeks?

By the Rule,

5 men : 16 men $\Big\}$ ::£18 15s.
12 wks. : 20 wks.

16 men will earn *more* money tha̶n̶ ̶5̶ men in a *given* ti̶m̶e̶, ̶a̶n̶d̶ in wks. *more* money will be earned than in 12 w̶k̶s̶.̶ ̶b̶y̶ ̶a̶ given no. *of men.*

∴ 5 × 12 : 16 × 20 :: 375s. : no. of shillings req^d. ;

∴ no. of shillings req^d. = $\dfrac{16 \times 20 \times 375}{5 \times 12}$ = 2000s. = £100.

ARITHMETIC.

Ex. 2. If 16 horses eat 56 bus. of corn in 32 days, in how many days will 8 horses eat 84 bus.?

8 horses : 16 horses }
56 bus. : 84 bus.   } :: 32 days

$\therefore$ no. days req^d. $= \dfrac{16 \times 84 \times 32}{8 \times 56} = 96.$

A given no. of bus. will last 8 horses *more* days than 16 horses; 84 bus. will last *a given no.* of horses *more* days than 56 bus.

Ex. 3. If 15 pumps, working 8 hours a day, can raise 1260 tons of water in 7 days; how many pumps, working 12 hours a day, will be required to raise 7560 tons of water in 14 days?

12 hrs. : 8 hrs.   }
1260 tons : 7560 tons   } :: 15 pumps
14 days : 7 days   }

$\therefore$ no. of pumps req^d.

$= \dfrac{8 \times 7560 \times 7 \times 15}{12 \times 1260 \times 14} = 30.$

Fewer pumps work^g. 12 hrs. a day are req^d. to raise a *given weight* of water in *a given no.* of days than if they worked 8 hrs a day; *more* pumps are req^d. to raise 7560 tons than to raise 1260 tons in *a given no.* of days, work^g. *a given no.* of hrs. each day; *fewer* pumps are req^d., work^g. for 14 days *a given no.* of hrs. each day, to raise a *given weight* of water, than if they worked only for 7 days.

Ex. 4. If 25 men can perform a piece of work in 16 days working 12 hours a day, in what time will 20 men perform a similar piece of work 4 times as large, when they work only 8 hours a day?

Call the 1st piece of work 1, then the 2nd piece will = 4.

20 men : 25 men }
1 : 4   } :: 16 days.
8 hrs. : 12 hrs.}

$\therefore$ no. of days req^d.

$= \dfrac{25 \times 4 \times 12 \times 16}{20 \times 8} = 120.$

Ex. 5. A contractor engages to make a road 5½ mi. long in 160 days; but after employing 135 men upon it for 100 days, he finds that only 3 mi., 700 yards are completed; how many extra men must he employ in order to complete his contract?

5½ mi. − 3 mi., 700 yds. = 9680 yds. − 5980 yds. = 3700 yds.

5980 yds. : 3700 yds. }
60 days : 100 days.   } :: 135 men

$\therefore$ no. of men req^d.

$= \dfrac{3700 \times 100 \times 135}{5980 \times 60} = 139\tfrac{2\,9\,5}{2\,9\,9}$;

$\therefore$ 140 men must be employed, or 6 additional men.

DOUBLE RULE OF THREE.

Ex. LXVII.

(1) If 10 sacks of oats supply 12 horses for 4 weeks, how long will 15 sacks supply 9 horses?

(2) If 42 men finish a work in 36 days, how many will finish twice as great a work in 27 days?

(3) If 60 men in 36 days finish a work, in how many days will 135 men finish four times as great a work?

(4) If 104 tons carried 34 miles cost $87.36, what will 102 tons carried 122 miles cost?

(5) If a man with a capital of $100000 gain $2500 in 3 months, what sum will he gain with a capital of $1500000 in 7 months?

(6) If 21 cwt. be carried 40 miles for $2.80, how far ought 7 cwt. to be carried for $4.06?

(7) If 7 horses be kept 20 days for $70, what will it cost to keep 45 horses for 9 days?

(8) If 140 horses eat 560 bus. of oats in 16 days, how many horses may be kept for 24 days on 1200 bus. of oats?

(9) If with a capital of $5000 a person gains by trade $250 in 16 months, in how many months will he gain $625 with a capital of $2000?

(10) If a regiment of 1878 soldiers consume 702 qrs. of wheat in 336 days, how many qrs. will an army of 22536 men consume in 112 days?

(11) If 6 horses can plough $17\frac{1}{2}$ acres in 4 days, how much land can 24 horses plough in $2\frac{1}{4}$ days?

(12) If £240 be paid for bread for 49 persons for 20 mo., when wheat is 48s. a qr.; how long will £234 find bread for 91 persons, when wheat is £2. 16s. a qr.?

(13) If 100·8 lbs. of flour support 20 men for 3 days, how many men will 46·305 cwt. support for 7·35 weeks?

(14) If 26 men can reap a field of 85 ac. in 12 days, how many men will reap another similar field one-half the size of the 1st field in one-seventh part of the time?

(15) 3 men can do a piece of work in 6 days, if they work 10 hours a day; how long will it take 16 men to do twice the amount of work, when they are working at it 9 hours a day?

(16) If the wages of 25 men amount to £76. 13s. 4d. in 16 days, how many men must work 24 days to receive

£103. 10s., the daily wages of each of the latter being one-half that of each of those of the former?

(17) If 6664 men, on half rations, consume 357 qrs. of wheat in 57 days, how many qrs. of wheat will 798 men, on full rations, consume in 119 days?

(18) If the 16 cts. loaf weighs 3·35 lbs., when wheat is $1.14 a bus., what ought to be the price of wheat per bus., when 47·5 lbs. of bread cost $3.20.?

(19) If when wheat is $14.40 a qr., the 12 cts. loaf weighs 4 lbs., what should be the price of wheat per qr., when 25 lbs. of bread cost $37\frac{1}{2}$ cts.?

(20) If 4 men, each working 8 hrs. a day, take 11 days to pave a road 220 yds. long, and 35 ft. broad; how many days will 6 men, each working 12 hrs. a day, take to pave a road 175 yds. long, and 36 ft. broad?

(21) If 100 horses consume a stack of hay 20 ft. long, 11 ft., 3 in. broad, and 31 ft., 6 in. high, in 9 days, how long will a stack 18 ft. long, 5 ft. broad, and 14 ft. high supply 80 horses?

(22) If 3 men can dig a ditch 105 yds. long, 4 ft. deep, and 5 ft. wide in 10 days, how long will it take 5 men to dig a ditch 175 yds. long, $4\frac{1}{2}$ ft. deep, and 6 ft. wide?

(23) If the 8 cts. loaf weighs 1 lb. 11 oz. 12 drs. when wheat is $1·80 per bu., what ought the 12 cts. loaf to weigh when wheat is $1.26 per bus.?

(24) If 5 horses require as much corn as 8 ponies, and 15 qrs. last 12 ponies for 64 days, how many horses may be kept 48 days for £41. 5s. when corn is 22s. a qr.?

(25) A contractor agrees to execute a certain piece of work in a certain time. He employs 55 men, who work 9 hrs. daily. When $\frac{3}{5}$ths of the time is expired, he finds that only $\frac{2}{5}$ths of the work is done. How many men must he employ during the remaining part of the time, working 11 hrs. daily, in order that he may fulfil his contract?

(26) If 5 pumps, each having a length of stroke of 3 feet, working 15 hours a day for 5 days, empty the water out of a mine; what must be the length of stroke of each of 15 pumps which, working 10 hours a day for 12 days, would empty the same mine, the strokes of the former set of pumps being performed 4 times as fast as those of the latter?

PRACTICE.

120. An *Aliquot* part of a number is such a part as, when taken a certain number of times, will exactly make up that number.

Thus, 4 is an aliquot part of 12; 6s. of 18s.

TABLES OF ALIQUOT PARTS.

Parts of a cwt. (100 lbs.)

50 lbs. or 2 qrs.	= $\frac{1}{2}$ cwt.	
25 lbs. or 1 qr.	= $\frac{1}{4}$	"
20 lbs.	= $\frac{1}{5}$	"
10 lbs.	= $\frac{1}{10}$	"
5 lbs.	= $\frac{1}{20}$	"

Note. The parts of a $ the same as of the cwt. (100 lbs).

Parts of a cwt. (112 lbs.)

56 lbs. or 2 qrs.	= $\frac{1}{2}$ cwt.	
28 lbs. or 1 qr.	= $\frac{1}{4}$	"
16 lbs.	= $\frac{1}{7}$	"
14 lbs.	= $\frac{1}{8}$	"
7 lbs.	= $\frac{1}{16}$	"
4 lbs.	= $\frac{1}{28}$	"
2 lbs.	= $\frac{1}{56}$	"

Parts of a £1.

10s.	= $\frac{1}{2}$ £1.	
6s. 8d.	= $\frac{1}{3}$	"
5s.	= $\frac{1}{4}$	"
4s.	= $\frac{1}{5}$	"
3s. 4d.	= $\frac{1}{6}$	"
2s. 6d.	= $\frac{1}{8}$	"
2s.	= $\frac{1}{10}$	"
1s. 8d.	= $\frac{1}{12}$	"
1s. 4d.	= $\frac{1}{15}$	"
1s. 3d.	= $\frac{1}{16}$	"
1s.	= $\frac{1}{20}$	"

Parts of a shilling.

6d.	= $\frac{1}{2}$ of 1s.	
4d.	= $\frac{1}{3}$	"
3d.	= $\frac{1}{4}$	"
2d.	= $\frac{1}{6}$	"
1$\frac{1}{2}$d.	= $\frac{1}{8}$	"
1d.	= $\frac{1}{12}$	"
$\frac{3}{4}$d.	= $\frac{1}{16}$	"
$\frac{1}{2}$d.	= $\frac{1}{24}$	"
$\frac{1}{4}$d.	= $\frac{1}{48}$	"

Note. In working examples in Practice, the above tables will often have to be varied; the knowledge, which the scholar now has, will render him expert in taking such aliquot parts as he may require in any particular example.

121. Practice is a short method of finding the value of any number of articles by means of *Aliquot Parts*, when the value of a unit of any denomination is given. Practice may be divided into Simple and Compound.

SIMPLE PRACTICE.

122. In this case the given number is expressed in the same denomination as the unit whose value is given; as, for instance, 27 bushels at $1.30 per bushel.

ARITHMETIC.

The Rule for Simple **Practice** will be easily shewn by the following examples.

Ex. 1. Find the **value of 1**296 things at 16s. 10½d. each.
The method **of working** such an example is the following:
　　If the cost of the things be £1 each;
　　　then the total cost = £1296.

∴ cost at
　　　　　　　　　　　　　　　　　　　　£.　　 s.　 d.
　10s. 0d. each = ½ of the above sum. .... = 648 . 0 . 0
　5s. 0d. each = ½ the cost at 10s. each.. = 324 . 0 . 0
　1s. 3d. each = ¼ the cost at 5s. each. .. = 81 . 0 . 0
　0s. 7½d. each = ½ the cost at 1s. 3d. each = 40 . 10 . 0

∴ by adding up the vertical columns,
cost at 16s. 10½d. each　　　　　= £1093 . 10 . 0

The operation is usually written thus:

　　　　　　　　　　　£.　　s.　d.
10s. = ½ of £1.　　1296 . 0 . 0 = cost at £1 **each**.
　5s. = ½ of 10s.　　648 . 0 . 0 = cost at 10s. each.
1s. 3d. = ¼ of 5s.　 324 . 0 . 0 = **cost at** 5s. each.
7½d. = ½ of 1s. 3d.　 81 . 0 . 0 = **cost at** 1s. 3d. each.
　　　　　　　　　　　 40 . 10 . 0 = cost at 7½d. each.
　　　　　　　　　£1093 . 10 . 0 = cost at 16s. 10½d. each.

Note. The student must use his own judgment in selecting the most convenient 'aliquot' parts; taking care that the sum of those taken make up the *given price of the unit.*

Ex. 2. Find the **value of 825 bushels of wheat at** $1.30 per bus.
　If 1 bus. cost $1, cost of 825 bus. = $825 at $1 each.

　　　　　　　　　　　$825.00 = value at $1 each.
20 cts. = ⅕ of $1.　　　165.00 = value at 20 cts. each.
10 cts. = ½ of 20 cts.　　82.50 = value at 10 cts. each.
　　　　　　　　　　　$1072.50 = **value at** $1.30 each.

Ex. LXVIII.

Find the value of,
(1) 75 at $2.25.　　　　(2) 105 at $1.50.
(3) 910 at $1.75.　　　 (4) 876 at $2.20.
(5) 1075 at $3.**25.**　　(6) 1278 at $1.87½.
(7) 397 at £1. 1s.　　　(8) 250 at £2. 8s.
(9) 1324 at $3.75.　　 (10) 2678 at £2. 7s. 6d.
(11) 973 at 16s. 8½d.　 (12) 236 at £7. 5s. 11¼l.

PRACTICE.

(13) 9078 at £8. 13s. 8½d.   (14) 15739 at £9. 17s. 9¾d.
(15) 27835 at $9.62½.   (16) 37832 at $18.90.
(17) A bankrupt whose debts amount to $250215 pays 29 cts. in the dollar; what are his effects worth?
(18) A gentleman's gross income is $12815, his rates and taxes amount to 25 cts. in the $, find his net income.
(19) What will be the loss on a debt of £4970, if a dividend of 8s. 10½d. in the £ be paid?
(20) What will be the total cost of 83½ yds. of calico @ 11½d. per yd., of 57¾ yds. of flannel @ 1s. 10d. a yd., and of 118 yds. of ribbon @ 9¼d. a yd.?

COMPOUND PRACTICE.

123. In this case the given number is not wholly expressed in the same denomination as the unit whose value is given; as for instance, 1 cwt. 2 qrs., 14 lbs. at $10.24 per cwt.

The Rule for *Compound Practice* will be easily shewn from the following examples.

Ex. 1. Find the value of 60 cwt., 3 qrs., 5 lbs. of sugar @ $8.50 per cwt.

The method of working such an example is the following:
The value of 1 cwt. of sugar being $8.50;

∴ value of 60 cwt. = ($8.50 × 60) = $510.00
2 qrs. = ½ (value of 1 cwt.)
   = ½ ($8.50) = $4.25
1 qr. = ½ (value of 2 qrs.)
   = ½ ($4.25) = $2.12½
5 lbs. = ⅕ (value of 1 qr.)
   = ⅕ ($2.12½) = $0.42½

Therefore adding up the vertical columns,
value of 60 cwt. 3 qrs., 5 lbs.   = $516.80

The operation is usually written thus:

2 qrs. = ½ cwt. | $8.50 = value of 1 cwt.
 | 10
 | 85.00 = value of 10 cwt.
 | 6
 | 510.00 = value of 60 cwt.
 | 4.25 = value of 2 qrs.
1 qr. = ½ of 2 qrs. | 2.12½ = value of 1 qr.
5 lbs. = ⅕ of 1 qr. | .42½ = value of 5 lbs.
 | $516.80 = value of 60 cwt., 3 qrs., 5 lbs.

ARITHMETIC.

value of 319 cwt., 3 qrs., 16 lbs., **at £2.**

	£	s.	d.	
	2 .	12 .	6	= value of 1 cwt.
			10	
	26 .	5 .	0	= value of 10 cwt.
			4	
	105 .	0 .	0	= value of 40 cwt.
			8	
	840 .	0 .	0	= value of 320 cwt.
subtracting	2 .	12 .	6	= value of 1 cwt.
	837 .	7 .	6	= value of 319 cwt.
	1 .	6 .	3	= value of 2 qrs.
1 qr. = ½ of 2 qrs.	0 .	13 .	1½	= value of 1 qr.
14 lbs. = ½ of 1 qr.	0 .	6 .	6¾	= value of 14 lbs.
2 lbs. = ⅐ of 14 lbs.	0 .	0 .	11¼	= value of 2 lbs.
	£839 .	14 .	4½	= value of 319 cwt., 3 qrs., 16 lbs.

Ex. LXIX.

Find the **value of**

(1) 55 bus., 25 lbs. **wheat** @ $1.20 per bushel.

(2) 16 cwt., 2 qrs., 20 lbs. of sugar @ 10 cts. per lb.

(3) 96 ac., 2 ro., 10 per. at $15.50 per ac.

(4) 2 lbs., 8 oz., 13 dwt. at 7s. 1d. per oz.

(5) 15 yds., 2 ft., 7 in. at 12s. 6d. per yd.

(6) 28 sq. yds., 7 ft., 110 in. at £1. 7s. per sq. ft.

(7) 11 mls., 3 fur., 55 yds., at $11000 per mile.

(8) What is the value of 5 tubs of butter, each of 2 of them containing 57½ lbs., and each of **the rest** 73¾ lbs., at $25 per cwt.?

(9) What will 3460 ft. of timber cost at $5 per 100 ft.?

(10) What will 24650 bricks cost at $4 per 1000?

(11) What will 46390 ft. lumber cost at $10.25 per 1000 ft.?

Find the amount of each of the following bills:

(12) 17⅞ yds. calico at 19½ cts. a yd., 35⅜ yds. flannel at 55½ cts. a yd., 96⅓ yds. sheeting at 70½ cts. a yd., 104¾ yds of Holland at 32½ cts. a yd., 12⅞ yds. of ribbon at 17¼ cts. a yd.

(13) 25¾ lbs. of beef at 12½ cts. a lb., 19½ veal at 11 cts.

a lb., 35⅞ lbs. of pork at 8½ cts. a lb., 17½ lbs. lamb at 6½ cts. a lb.

(14) 17⅔ lbs. crushed sugar at 12½ cts. a lb., 18¾ lbs. cheese at 17½ cts. a lb., 5₁⁵⁄₁₃ lbs. of tea at 75 cts. a lb., 10⅞ lbs. coffee at 40 cts. a lb., 7¾ lbs. honey at 25 cts. a lb.

Note 1. The scholar should bring the last three questions in the form of a bill, to the master.

INTEREST.

124. INTEREST (Int.) is the sum of money paid for the loan or use of some other sum of money, lent for a certain time at a fixed rate; generally at so much for each $100 for one year.

The money lent is called THE PRINCIPAL.

The int. of $100 for a year is called THE RATE PER CENT.

The principal + the interest is called the AMOUNT.

Interest is divided into Simple and Compound. When interest is reckoned only on the principal or sum lent, it is SIMPLE INTEREST.

When the interest at the end of the first period, instead of being paid by the borrower, is retained by him and added as principal to the former principal, interest being calculated on the new principal for the next period, and this interest again, instead of being paid, is retained and added on to the last principal for a new principal, and so on; it is COMPOUND INTEREST.

SIMPLE INTEREST.

125. *To find the interest of a given sum of money at a given rate per cent. for a year.*

RULE. Multiply the principal by the rate per cent., and divide the product by 100.

Note 2. The int. for any given number of years will be found by multiplying the int. for 1 year, by the number of years; and the int. for any part of a year may be found from the int. for 1 year, by Practice, or by the Rule of Three.

Note 3. If the interest has to be calculated from one given day to another, as for instance from the 30th of Jan. to the 7th of Feb., the 30th of Jan. must be left out in the calculation, and the 7th of Feb. must be taken into account, for the borrower will not have had the use of the money for one day till the 31st of Jan.

Note 4. If the amount be required, the int. has first to be found for the given time, and the principal has then to be added to it.

Ex. 1. Find the simple int. of $250 for one year, at 9 per cent. per annum.

By the Rule,   or by the Rule of Three.

$250 $100 : $250 :: $9 : Int. req^d.,
   9
─────
$22.50
∴ Int. = $22.50   ∴ Int. req^d. = $\dfrac{250 \times 9}{100}$ = $22.50.

Ex. 2. Find the amount of £1376. 11s. 3d. at 4¾ per cent. from Apr. 6 to Aug. 30.

```
  £.    s.   d.              £.    s.   d.
1376 . 11 .  3             1376 . 11 .  3
          4¾                          3
─────────────             ─────────────
5506 .  5 .  0           4)4129 . 13 .  9
1032 .  8 . 5½             £1032 .  8 . 5½
─────────────
£6538 . 13 . 5¼
       20
     ─────
   s. 7·73
       12
     ─────
   d. 8·8125   since 5¼d. = 5·25d.
```

∴ Int for 1 yr. = £65. 7s. 8·8125d.

No. of days from Apr. 6 to Aug. 30 = 24 + 31 + 30 + 31 + 30 = 146 ;

∴ 365 days : 146 days :: £65. 7s. 8·8125d. : int. req^d.
  or 5 : 2 :: £65. 7s. 8·8125d. : int. req^d.

∴ int. req^d. = ⅖ of £65. 7s. 8·8125d. = £26. 3s. 1·125d. ;

∴ Am^t. = £1376. 11s. 3d. + £26. 3s. 1·125d. = £1402. 14s. 4·125d.

Note. Since £1276. 11s. 3d. = £1376.5625, and 4¾ = 4·75, we might have found the int. thus : int. = £$\left(\dfrac{1376 \cdot 5625 \times 4 \cdot 75}{100}\right)$

= £65·38671875.

Ex. LXX.

Find the Simple Int. and also the Amt. of

(1) $217.25 for 1 year at 8 per cent. per ann^m.
(2) $217.25 for 2 yrs. at 8 per cent............
(3) $527.37½ for 3 yrs. at 7................
(4) $93.50 for 2 yrs. at 6..................
(5) $75.75 for 2½ yrs at 7..................
(6) £62. 18s. 9½d. for 3½ yrs. at 8...........

COMPOUND INTEREST.

(7) $1075.75 for 4¼ yrs. at 8 per cent. per ann^m.
(8) $684 for 5 yrs. 8 mo., at 8 ....
(9) £7500 from **May 5 to Oct. 26, at 3⅛** ....
(10) £4865. 11s. 5d. from Jan. 1 to Aug. 28, 1868, at 5⅜.
(11) In what time will $672 at 8 per cent. simp. int. amount to $994.56 ?
(12) At what rate per cent., simp. int., will the int. on $816 amount to $346.80 in 5 yrs. ?
(13) What sum of money will amount to £138. 2s. 6d. in 15 mo. at 5 per cent. per ann^m., simp. int. ?
(14) If £1 = 10 florins = 100 cents = 1000 mills, find the simp. int. on £578. 3 fl. 1 c. 2½ m. for 2¼ yrs. at 2½ per cent.
(15) At what rate per cent., simp. int., will $2293.75 double itself in 25 yrs. ?

COMPOUND INTEREST.

126. *To find the Compound Interest of a given sum of money at a given rate per cent. for any number of years.*

RULE. At the end of each year add the interest of that year, found by (Art. 116), to the principal at the beginning of it; this will be the principal for the next year; proceed in the same way as far as may be required by the question. Add together the interests so arising in the several years, and the result will be the compound interest for the given period.

Ex. 1. Find the Comp. Int. and Am^t. of $600 for 3 yrs. at 8 per cent. per ann.

$600
  8
―――――
$48.00   Int. for 1st yr.
∴ $648   Prin^l. for 2nd yr.
    8
―――――
$51.84   Int. for 2nd yr.
∴ $699.84   Prin^l. for 3rd yr.
    8
―――――
$55·9872   Int. for 3rd yr.

∴ Comp^d int. = $55·9872 + $51.84 + $48 = $155·8272.
Am^t. $600 + $155·8272 = $755·8272.

Ex. 2. Find, **working with decimals, the comp. int. and** am^t. of £690 for 2 yrs. at 4½ per cent. per ann.

$$\begin{array}{r}£\\690\\4\tfrac{1}{2}=\quad 4\cdot 5\\\hline 3450\\2760\\\hline £31\cdot 050=\text{Int. for 1st yr.}\\£690\\\hline £721\cdot 050=\text{Prin}^l.\text{ for 2nd yr.}\\4\cdot 5\\\hline 360525\\288420\\\hline £32\cdot 44725=\text{Int. for 2nd yr.}\\£721\cdot 05\\\hline £753\cdot 49725=\text{Prin}^l.\text{ for 3rd yr. or amount req}^d\\20\\\hline 9\cdot 94500s.\\12\\\hline 11\cdot 340d.\\4\\\hline 1\cdot 36q.\end{array}$$

∴ am^t. = £753. 9s. 11¼d. nearly, and Int. = £753. 9s. 11¼d., nearly – £690 = £63. 9s. 11¼d. nearly.

Note 1. It is customary, if the comp^d. int. be required for any number of entire yrs. and a part of a yr. (for instance for 5¾ yrs.) to find the comp^d. int. for the 6th yr., and then take ¾ths of the last int. for the ¾ths of the 6th yr.

Note 2. If the int. be payable half-yearly or quarterly, it is clear that the comp^d. int. of a given sum for a given time will be greater as the length of each given period is less; the simp. int. will not be affected by the length of each period.

Ex. LXXI.

Find the Compound Int. and Am^t. of

(1) $800 for **2** yrs. at **7** per cent. per annum.
(2) $742 for **3** yrs. at **8**....................
(3) $560 for **5** yrs. at **10** ....
(4) $308 for **1½** yrs. at **6**.................paid quarterly.
(5) $610 for **2** yrs. at **8** ....paid half-yearly.
(6) $1000 for **3** yrs. **at 7**................paid half-yearly.

PRESENT WORTH.

(7) Find the difference between the Amounts at simp. and comp. int. of (1) £880 for 2 ys. at 3½ per cent. (2) £1431. 5s. for three yrs. at 4 per cent.

PRESENT WORTH AND DISCOUNT.

127. A owes B $500, which is to be paid at the end of 9 months from the present time; it is clear that, if the debt be paid at once (int. being reckoned, we will suppose, at 8 per cent. per annum), B ought to receive a less sum of money than $500; in fact such a sum of money as will, being now put out at 8 per cent. int., amounts to $500 at the end of 9 months. The sum which B ought to receive *now* is called the Present Worth of the $500, due 9 months hence, and the sum to be deducted from the $500, in consequence of immediate payment, which is in fact the int. of the Present Worth, is called the Discount of the $500 paid 9 months before it is due; hence,

PRESENT WORTH is the actual worth at the present time of a sum of money due some time hence, at a given rate of interest.

DISCOUNT of a sum of money is the interest of the Present Worth of that sum, calculated from the present time to the time when the sum would be properly payable.

∴ Disc^t. = given sum *less* its P. Worth, and P. Worth = given sum *less* its Disc^t.

PRESENT WORTH.

128. RULE. Find the interest of $100 for the given time at the given rate per cent., and state thus:

$100 + its interest for the given time at the given rate per cent. : given sum :: $100 : present worth required.

Ex. 1. Find the present worth of $676, due 6 months hence, at 8 per cent. per annum.

By the Rule,

Int. on $100 for 6 mo. at 8 per cent. = $4.
∴ $104 : $676 :: $100 : P. Worth req^d.

hence P. Worth req^d. = $\dfrac{670 \times 100}{104}$ = $650.

Reason. $100 is the P. Worth of $104, due 6 mo. hence, ∴ we have the above statement by the Rule of Three.

Ex. 2. Find the present worth of £275. 6s. 8d. due 15 months hence at 4 per cent per annum.

Int. of £100 for 15 mo. at 4 per cent. $= \dfrac{15}{12}$ of £4 $=$ £5.

$\therefore$ £105 : £275⅓ :: £100 : P. Worth req^d.

$\therefore$ P. Worth req^d. $= £\dfrac{275\frac{1}{3} \times 100}{105} =$ £262. 4s. 5¼d. **nearly.**

DISCOUNT.

129. Rule. Find the interest of $100 for the given time at the given rate per cent., and state thus :

$100. + its interest for the given time at the given rate per cent. : given sum :: interest of $100 for the given time at the given rate per cent. : discount required.

Ex. 1. Find the discount of $250.75 due 17 months hence at 8 per cent per annum, simple interest.

By the Rule,

Int. of $100 for 17 mo. at 8 per cent. $= \dfrac{17}{12}$ of $8 $=$ $11⅓.

$\therefore$ $111⅓ : $250¾ :: $11⅓ : disc^t. req^d.

$\therefore$ disc^t. req^d. $= \$\dfrac{250\frac{3}{4} \times 11\frac{1}{3}}{111\frac{1}{3}} =$ $25.40⅞.

Reason. $11⅓ is the interest on $100 or the discount on $111⅓ for 17 mo. at 8 per cent., $\therefore$ we have the above statement by the Rule of Three.

130. In the discharge of a tradesman's bill before it has become due, it is usual to deduct interest instead of discount ; thus, if *B* contracts with *A* a debt of $100, *A* giving 12 months' credit, it is usual, if the interest of money be reckoned at 8 per cent. per annum, and the bill be discharged at once, for *A* to throw off $8, or for *A* to receive $92 instead of $100 ; but if *A* were to put out the $92 at 8 per cent. interest for 12 months it will not amount to $100 ; therefore such a proceeding is to the advantage of *B* : the sum of money which in strictness ought to have been deducted, was not $8, the interest on the whole debt, but $7.36, the interest on the present worth of the debt, *i. e.* the discount.

131. Bankers and Merchants in discounting bills calculate interest, instead of discount, on the sum drawn for in the bill, from the time of their discounting it to the time when it becomes due, adding THREE DAYS OF GRACE, which days are usually allowed after the time a bill is NOMINALLY due, be-

DISCOUNT.

fore it is LEGALLY due. When a bill is payable on demand, the days of grace are not allowed.

If a bill, without the days of grace, should appear to be due on the 31st of any month which contains less than 31 days, the last day of that month, and not the first day of the next, is considered as the day on which the bill is due. Thus a bill drawn on the 31st of Oct. at 4 months, would be really due, adding in the days of grace, on the 3rd of March. Bills which fall due on a Sunday, are paid on the previous Saturday.

Ex. A bill of £1000 is drawn on Feb. 16th, 1864, at 7 months' date: it is discounted on the 8th day of July at 5 per cent. What does the banker gain by the transaction?

The bill is *legally* due on Sept. 19; from July 8 to Sept. 19 are 73 days.

Int. of £1000 for 73 days $= £10$. Disc^t. $= £9.\ 18\tfrac{3}{101}s.$, $\therefore$ banker's gain $= £10 - £9.\ 18\tfrac{3}{101}s. = 1\tfrac{99}{101}s.$

Ex. LXXII.

Find the present worth of

(1) $216   due 1 yr. hence at 8 per ct. per ann. simp. int.
(2) $968   ......3 yr.......... 7 ....
(3) $1236   ......6 mo......... 6 ....
(4) $225.25   ......9 mo........ 10 ....
(5) $1057.50   ......$2\tfrac{1}{2}$ yrs........ 7 ....
(6) £161. 13s. $5\tfrac{1}{4}d$. $7\tfrac{1}{2}$ yrs........ $3\tfrac{1}{2}$ ....
(7) £193. 17s. $4\tfrac{1}{4}d$.   19 mo......... 5 ....
(8) £458. 8s. $9\tfrac{1}{4}d$.   31 days...... 5 ....

Find the Discount on

(9) $217   due 3 yrs. hence at 8 per ct. per ann. simp. int.
(10) $22100   .... $1\tfrac{1}{2}$ yrs......... 7 ....
(11) $2000   .... 6 mo........ 10 ....
(12) $1750   .... 9 mo......... 8 ....
(13) £345. 16s. 3d.   86 days...... 4 ....

(14) What is the difference between the true and mercantile discount on £549 for 32 days at 5 per cent. per annum?

(15) A bill for £450 drawn March 3, at 9 mo. date, is discounted by a banker on Oct. 22 at 5 per cent. Find his profit.

134   ARITHMETIC.

(16) From a bill of £3. 11s. 8d. due 18 mo. hence, a tradesman deducts 5s. ; which is the rate per cent. at which the true discount is calculated?

STOCKS.

132. If the 6 per cent. "Dominion of Canada" stock be quoted in the money market at 105½, the meaning is, that for $105½ of money a man can purchase $100 of such stock, for which he will receive a document which will entitle him to half-yearly payments of Interest or Dividends, as they are called, from the Government of the country, at the rate of 6 per cent. per annum on the stock held by him, until the Government choose to pay off the debt.

Similarly, if shares in any trading company, which were originally fixed at any given amount, say $100 each, be advertised in the share-market at 86, the meaning is, that for $86 of money *one* share can be obtained, and the holder of such share will receive a dividend at the end of each half-year upon the $100 share according to the state of the finances of the company.

STOCK may therefore be defined to be the capital of trading companies; or to be the money borrowed by our or any other Government, at so much per cent., to defray the expenses of the nation.

The amount of debt owing by the Government is called the NATIONAL DEBT, or the FUNDS. The Government reserves to itself the option of paying off the principal or debt at any future time, pledging itself however to pay the interest on it regularly at fixed periods in the mean time.

From a variety of causes the price of stock is continually varying. A fundholder can at any time sell his stock, and so convert it into money, and it will depend upon the price at which he disposes of it as compared with the price at which he bought it, whether he will gain or lose by the transaction.

Note 1. Purchases or sales of stock are made through Brokers, who generally charge $⅛, or 12½ cts. per cent. upon the stock bought or sold: so that, when stock is bought by any party, every $100 stock costs that party $⅛ more than the market-price of the stock: and when stock is sold, the seller gets $⅛ less for every $100 stock sold than the market-price.

Thus, the actual cost of $100 stock in the 3 per cents at

STOCKS.   135

$94\frac{1}{2}$, is $(94\frac{1}{2} + \frac{1}{8})$, or $94\frac{5}{8}$. The actual sum received for $100 stock in the 3 per cents. at $94\frac{1}{2}$, is $(94\frac{1}{2} - \frac{1}{8})$, or $94.

Unless the brokerage is mentioned, it need not be noticed in working examples in stocks.

Note. When $100 stock costs $100 in money, the stock is said to be at *par;* when $100 stock costs more than $100 in money, the stock is said to be at a *premium;* when $100 stock costs less than $100 money, the stock is said to be at a *discount.*

All Examples in Stocks depend on the principles of Proportion, and may therefore be worked by the Rule of Three.

Ex. 1. What sum of money will purchase $2600 6 per cent. stock at 93 ?

$100 stock (st) costs $93 in money ;

$\therefore$ $100 st. : £2600 st. :: $93 : req^d. sum ;

$\therefore$ req^d. sum = $\dfrac{2600 \times 93}{100}$ = $2418.

Ex. 2. Find the cost of £2353 3 per cent. Consols at $90\frac{3}{8}$ brokerage being $\frac{1}{8}$ per cent.

£100 st. costs £$(90\frac{3}{8} + \frac{1}{8})$, or $90\frac{1}{2}$;

$\therefore$ £100 st. : £2353 st. :: £$90\frac{1}{2}$: req^d. cost ;

$\therefore$ req^d. cost = £$\dfrac{2353 \times 90\frac{1}{2}}{100}$ = £2129. 9s. $3\frac{1}{4}d$. $\frac{3}{8}q$.

Ex. 3. A person who has $10000 Bank stock, sells out when it is at 35 per cent. premium; what amount of money does he receive, brokerage being $\frac{1}{8}$ per cent. ?

$100 st. sells for $ $\left(135 - \dfrac{1}{8}\right)$, or $134\frac{7}{8}$ money ;

$\therefore$ $100 st. : £10000 st. :: $134\frac{7}{8}$: req^d. amt. of money ;

$\therefore$ req^d. amt. = $ $\dfrac{10000 \times 134\frac{7}{8}}{100}$ = $13487.50.

Ex. 4. What incomes will $5500 at 7 per cent. stock, and $5500 invested in the 7 per cent. stock at $102\frac{2}{3}$, respectively produce?

1st, since every $100 stock gives $7 int. ; $\therefore$ income from $5500 of 7 per cent. stock = $$\dfrac{5500 \times 7}{100}$ = $385.

2nd, since $100 stock, which gives $7 int., costs $102\frac{2}{3}$; $\therefore$ every $102\frac{2}{3}$ give $7 int. ;

$\therefore$ $102\frac{2}{3}$: $5500 :: $7 req^d. income ;

$\therefore$ req^d. income = $ $\dfrac{5500 \times 7}{102\frac{2}{3}}$ = $375.

Ex. 5. One person buys £500 Consols at $90\frac{1}{2}$ and sells out at 93; another invests £500 in Consols at $90\frac{1}{2}$ and sells out at 93; what sum of money does each gain?

1st man gains $£(93 - 90\frac{1}{3})$, or $£2\frac{2}{3}$, on every £100 stock;
$\therefore$ his whole gain $= £(2\frac{2}{3} \times 5) = £13.\ 6s.\ 8d.$

2d man gains $£2\frac{2}{3}$ on every £100 stock, *i. e.* on every $£90\frac{1}{3}$ of his money which he invests:

$\therefore £90\frac{1}{3} : £500 :: £2\frac{2}{3} : $ whole gain ;

$\therefore$ whole gain $= £\dfrac{500 \times 2\frac{2}{3}}{90\frac{1}{3}} = £14.\ 15s.\ 2\frac{1}{4}d.$, nearly.

Ex. 6. A person invested some money in the 3 per cent. Consols when they were at 90, and some money when they were at 80; find the rate of interest he obtained in each case, and the advantage per cent. of the second purchase over the first.

£90 : £100 :: £3 : rate per cent. in 1st case,
£80 : £100 :: £3 : rate per cent. in 2d case,

$\therefore$ rate per cent. in 1st case $= £\dfrac{100 \times 3}{90} = £3.\ 6s.\ 8d.$;

$\therefore$ .... 2nd.... $= £\left(\dfrac{100 \times 3}{80}\right) = £3.\ 15s$;

$\therefore$ advantage $= £3.\ 15s. - £3.\ 6s.\ 8d. = 8s.\ 4d.$

Ex. 7. A person invests £1037. 10s. in the 3 per cents. at 83; the funds rise 1 per cent.; he then transfers his capital to the 4 per cents at 96: find the alteration in his income.

£83 : £1037. 10s. :: £100 : quantity of 3 per cent. st.;

$\therefore$ quantity of 3 per cent st. bought $= £\dfrac{1037\frac{1}{2} \times 100}{80} = £1250.$

The funds have risen 1 per cent. therefore to transfer £1250 stock from the funds at 84 to the funds at 96,

£96 : £84 :: £1250 stock : quantity of 4 per cent. stock, (since the higher the price of the stock the less will be the amount purchased);

$\therefore$ quantity of 4 per cent. stock $= £\dfrac{1250 \times 84}{96} = £1093.\ 15s.$

1st Income $= £\dfrac{1250 \times 3}{100} = £37.\ 10s.$

2nd Income $= £\dfrac{1093\frac{3}{4} \times 4}{100} = £43.\ 15s.$;

$\therefore$ alteration in income $= £43\ 15s. - £37\ 10s. = £6.\ 5s.$

Ex. LXXIII.

(1) Find amount of Bank of Montreal stock purchased by investing $527.25 at $126\frac{1}{2}$, the stock yielding 8 per cent., per annum interest?

(2) Bank of Toronto stock being at $102\frac{1}{2}$, how much can be purchased for $800?

(3) Find the value of $1556 Royal Canadian Bank stock at 98.

(4) Royal Canadian Bank stock being at 1 per cent. discount, I invest $525.50; find my income therefrom; the Bank's dividends being 7 per cent. per annum.

(5) Montreal Bank stock being at $125\frac{3}{8}$, and paying yearly dividends of $7\frac{1}{2}$ per cent.; how much money must be invested in order to secure an annual income of $900, allowing $\frac{1}{8}$ per cent. for brokerage?

(6) Upper Canada Bank bills are at 65; how much money could a person obtain for $2140 of such Bank bills?

(7) If a man invest £666. 8s. 4d. in the 3 per cents. at $90\frac{3}{4}$, (1) what half-yearly interest will he obtain after deducting an inc°. tax of 4d. in the £? (2) What rate per cent. will he get for the money invested?

(8) What rate per cent. per annum does a person receive for his money, who invests in Bank of Montreal stock at 136; the stock yielding half-yearly dividends of 4 per cent?

(9) Which would be the better investment, Bank of Montreal stock at 136, or Bank of Toronto stock at 104; half-yearly dividends being 4 and $3\frac{3}{4}$ per cent. respectively?

(10) If a person lay out £4650 in the $3\frac{1}{2}$ per cents. when they are at 7 per cent. discount, what will be his loss of property by the stocks falling $\frac{1}{2}$ per cent.?

(11) If a person were to transfer £29000 stock, from the $3\frac{1}{2}$ per cents. at 99 to the 3 per cents. at $90\frac{3}{8}$, what difference would it make in his income?

(12) A person invests $2000 in Bank of Toronto stock at 115, shortly afterwards he sells when the stock rose to 123. Find his gain?

(13) If the **3 per** cents. are at 95, and Government offer to receive tenders for a loan of £5016000, the lender to receive five millions in the 3 per cents., together with a certain sum in the $3\frac{1}{2}$ per cents.; what sum in the $3\frac{1}{2}$ per cents. ought the lender to accept?

(14) A man sells out of the $3\frac{1}{2}$ per cents. at $93\frac{1}{2}$ and realizes £18700: if he invest one-fifth of the produce in the 4

per cents. at 96, and the remainder in **the 3 per cents. at 90 ;** find the alteration in his income.

(15) A person **invests £5460** in the 3 per cents. at 91; he **sells out** £2000 **stock when they** have risen to 93½, **and the remainder when they have fallen to 85; he** then invests the produce in the 4½ **per cents. at 102.** What is the **difference in his income?**

(16) A person **has an income of £350 from money invested in the new 3** per cents., he **sells out at 87⅜, and invests in the India 5 per cents. at** 104⅞. **How will his income be affected,** ⅛th per cent. being allowed for brokerage?

APPLICATIONS OF THE TERM "PER CENT."

133. There are many other cases in which the term PER CENT. occurs besides those already mentioned; we will mention certain cases, and give examples in each by way of illustration.

COMMISSION **is the sum of money which** a **merchant** charges for buying or selling goods for another.

BROKERAGE is **of the same nature as Commission, but has** relation to money **transactions, rather than dealings** in goods or merchandise.

INSURANCE is a **contract, by which one party,** on being paid a certain sum or *Premium* by another party on property, **which is subject to risk,** undertakes, **in case of loss, to** make good **to the owner the value of that property. The document which expresses the contract is called** *the Policy of Insurance.*

LIFE ASSURANCE is **a contract for** the payment of a certain **sum of money on the death of** a person, in consideration **of an annual premium to be** continued **during** the life of *the Assured,* **or for a** certain number of **years.**

Questions on Commission, Brokerage, and Insurance, **these charges being** usually made **at** so **much** per cent., amount to **the same thing as** finding the interest on a given sum of money **at** a given rate **for** 1 yr., **and may therefore be** worked **by the** Rule for **Simple Int. or by the Rule of** Three.

Ex. 1. What is **the brokerage on the purchase of $4300** 6 per cents. stock at ⅛ per cent. ?

$$100 : 4300 :: \tfrac{1}{8} : \text{brok}^e. \text{ req}^d. \;\therefore\; \text{brok}^e. \text{ req}^d. = \$\frac{4300 \times \tfrac{1}{8}}{100} = \$5.37\tfrac{1}{2}$$

PERCENTAGE.

Ex. 2. What is the premium on a policy of insurance for £9626. 11s. 3d., at £2. 12s. per cent.?
    £100 : £9626. 11s. 3d. :: £2. 12s. : premium req^d.
    ∴ premium req^d. = £$\dfrac{9626\tfrac{9}{16} \times 2\tfrac{3}{5}}{100}$ = £250. 5s. 9$\tfrac{3}{4}$d.

Ex. 3. What is the annual cost of insuring property to the amount of $1600, the premium being $1.50 per cent.?
    $100 : 1600 :: 1.50 : ann^l. cost; ∴ ann^l. cost = $1.50 × 16 = $24.

134. All questions which relate to gain or loss in mercantile transactions fall under the head of PROFIT and LOSS.

Tradesmen measure their Profit or Loss by the actual amount gained or lost, or by the amount gained or lost on every $100 of the capital they invest.

Ex. 4. If tea be bought at 84 cts. per lb., and sold at 93 cts. per lb., find the gain per cent.
    (93 cts. − 84 cts.) = 9 cts.; ∴ gain on 84 cts. = 9 cts.
    ∴ 84 cts. : $100 :: 9 cts. : gain per cent. ;
    ∴ gain per cent. = $\dfrac{100 \times 9}{84}$ cts. = 10.71\tfrac{4}{7}$.

Ex. 5. If tea be bought at 93 cts. per lb. and sold at 84 cts. per lb., find the loss per cent.

In this case 9 cts. is lost on 93 cts.,
    ∴ 93 cts. : $100 :: 9 cts. : loss per cent.
whence loss per cent. = 9.67\tfrac{27}{31}$.

Ex. 6. By selling cheese at £3. 13s. 6d. a cwt. a grocer realized a profit of 22$\tfrac{1}{2}$ per cent., what did it cost him per cwt.?
He sells cheese for which he gave £100 for £122$\tfrac{1}{2}$.
    ∴ £122$\tfrac{1}{2}$: £3. 13s. 6d. or £3$\tfrac{27}{40}$ :: £100 : prime cost per cwt ;
    ∴ prime cost per cwt. = £$\dfrac{3\tfrac{27}{40} \times 100}{122\tfrac{1}{2}}$ = £3.

Ex 7. By selling cheese at £3. 13s 6d. a cwt. a grocer lost 22$\tfrac{1}{2}$ per cent., find the prime cost of the cheese per cwt.
In this case he sells cheese, for which he gave £100, for (£100 − £22$\tfrac{1}{2}$), or for £77$\tfrac{1}{2}$.
    ∴ £77$\tfrac{1}{2}$: £3$\tfrac{27}{40}$ :: £100 : prime cost of cheese per cwt. ;
    ∴ prime cost per cwt. = £$\dfrac{3\tfrac{27}{40} \times 100}{77\tfrac{1}{2}}$ = £4. 14s. 10$\tfrac{2}{31}$d.

Ex. 8. By selling sheep for $19 the seller loses 5 per cent. on his outlay; what would have been his loss or gain per cent. if he had sold the sheep for $23.75 ?

1st. $95 : $19 :: $100 :$ prime cost of sheep,
∴ prime cost of sheep $= \$20$.

2nd. $\$20 : \$100 :: \$3.75 :$ **gain per** cent., if the sheep be sold for $23.75;

$$\therefore \text{ gain per cent.} = \$ \frac{100 \times 3\frac{3}{4}}{20} = \$18.75.$$

This sum might have been worked thus,
$\$19 : \$23\frac{3}{4} :: \$95$, *i. e.* what $100 will realize if the sheep be sold for $19: what $100 will realize if the sheep be sold for $23\frac{3}{4}$.

$$\therefore \$100, \text{ if sheep sold for } \$23\frac{3}{4} \text{ will realize } \$ \frac{95 \times 23\frac{3}{4}}{19}, \text{ or } \$118\frac{3}{4};$$

∴ gain per cent. $= \$118\frac{3}{4} - \$100 = \$18\frac{3}{4} = \$18.75$.

135. Tables respecting the increase or decrease of Population, &c., are constructed with reference to the increase or decrease on every 100 of such population; Education returns are constructed in the same way; and so are other *Statistical Tables.*

Ex. 9. In 1852 the population of the County of Wellington was 26796, in 1861 it was 49200; find the increase per cent. $49200 - 26796 = 22404$; ∴ $26796 : 100 :: 22404 :$ incr^e. per cent.

$$\therefore \text{ increase per cent.} = \frac{2240400}{26796} = 83 \cdot 609 \ldots . \text{per cent.}$$

Ex. 10. Between the years 1841 and 1851 the population of England increased 14·2 per cent. In 1851 it was 21121290, what was it in 1841?

For every 100 persons in 1841 there were 114·2 in 1851;
∴ $114 \cdot 2 : 21121290 :: 100 :$ population in 1841;

$$\therefore \text{ population in 1841} = \frac{21121290 \times 100}{114 \cdot 2} = 18495000.$$

Ex. 11. If of a regiment of 750 men, 26 per cent. are in hospital, 32 per cent. in trenches, and the rest in camp, how many are in hospital, trenches, and camp, respectively.

$$100 : 750 :: 26 : \text{no. in hosp}^l.; \therefore \text{ no. in hosp}^l. = \frac{750 \times 26}{100} = 195.$$

$$100 : 750 :: 32 : \text{no. in tren}^{hs}.; \therefore \text{ no. in tren}^{hs}. = \frac{750 \times 32}{100} = 240.$$

∴ number in camp $= 750 - (195 + 240) = 315$.

Ex. 12. The percentage of children who are learning to

write is 65 in a school of 60 children, and 78 in another school of 70, what is the percentage in the two schools together?

In the 1st school,

$100 : 60 :: 65 :$ no. who write ; $\therefore$ no. who write $= \dfrac{60 \times 65}{100} = 39.$

In the 2nd school,

$100 : 70 :: 78 :$ no. who write ; $\therefore$ no. who write $= \dfrac{70 \times 78}{100} = 54\frac{3}{5}.$

$\therefore$ in a school of 130, there are $93\frac{3}{5}$ who write ;

$\therefore 130 : 100 :: 93\frac{3}{5} :$ percent. req^d. ; $\therefore$ percent. req^d.

$= \dfrac{100 \times 93\frac{3}{5}}{130} = 72.$

Ex. LXXIV.

(1) What will be the broker's commission on the purchase of $4300 6 per cents. at 90½, at ⅛ per cent. ?

(2) What is the premium on a policy of insurance for $9626.55 at $2.60 per cent. ?

(3) The commission on the purchase of $1560 Dominion stock at 104 amounted to $4.60, what was the rate per cent. ?

(4) For what sum would the life of a person aged 23 be insured by the annual payment of $45.60, the premium for that age being $2.40 per cent. ?

(5) A draper at Hamilton buys 25 pieces of calico, each containing 36 yds., for £32. 16s. 3d. ; the carriage costs him 6s. 3d. ; (1) What will he gain by selling the calico at 10½d. a yard? (2) What will he gain per cent. ?

(6) A merchant bought 1280 bus. of wheat at $1.20 a bu., the expenses of carriage, &c., averaged 3¾ cts. a bu. ; he sold the wheat at $1.40 a bu. (1) What was his gain ? (2) What was his gain per cent. ? (3) At what price a bu. should he have sold the wheat in order to gain $400 ?

(7) (1) A man buys a pig for 6s. 8d., and sells it for 7s. 4d.; find his gain per cent. (2) What would have been the loss per cent. had he bought the pig at 7s. 4d. and sold it at 6s. 8d. ?

(8) Tea is bought at $96 per cwt., at what price per lb. must it be sold to gain 25 per cent. ?

(9) Sugar is bought at $6 per cwt., what will be the gain per cent. if it be sold at 10 cts. per lb. ?

(10) At what price must a yd. of cloth be sold, which cost 4s. 8d., so as to gain 12½ per cent. ?

142 ARITHMETIC.

(11) If a yd of cloth, sold at 4s. 8d., give a profit of $12\frac{1}{2}$ per cent.; find the prime cost.

(12) A grocer buys 40 lbs. of tea at 84 cts., 44 lbs. at 93 cts., and 55 lbs. at $1.08; and sells the mixture for $188.16, what is his gain per cent.?

(13) A grocer mixes 26 lbs. of tea at 5s. 3d., 32 lbs. at 5s. 7d., and 36 lbs. at 6s. 1d.; at what rate per lb. must he sell the mixture in order to gain 40 per cent. on his outlay?

(14) If I sell for 15s. I *lose* 10 per cent., what must I sell at to *gain* 10 per cent.?

(15) A person buys a certain number of eggs and sells them again at such a price, that 11 are sold for the money 18 cost him. Find his gain per cent.

(16) A boy sells another boy a cricket-bat for $1.56, gaining thereby 30 per cent.; what did it cost him?

APPLICATIONS OF THE TERM "AVERAGE."

136. Questions are often given, in which the term "Average" occurs; two such examples will be worked by way of illustration, and others subjoined for practice.

Ex. 1. A gentleman in each of the following years expended the following sums: in 1845 $650, in 1846 $675, in 1847 $680, in 1848 $690, in 1849 $700, in 1850 $715, in 1851 $790. Find his average yearly expenditure.

The object is to find that fixed sum which he might have spent in each of the seven years, so that his total expenditure in that case might be the same as his total expenditure was in the above question.

Adding the various sums together we find that the total expenditure amounted to $4900; this sum divided by 7 gives $700 as the average yearly expenditure.

Ex. 2. In a school of 27 boys, 1 of the boys is of the age of 17 years, 2 of 16, 4 of $15\frac{1}{2}$, 1 of $14\frac{3}{4}$, 2 of $14\frac{1}{2}$, 5 of $13\frac{3}{4}$, 10 of $12\frac{1}{4}$, and 2 of 10; find the average age of the boys.

The object is to find, what must be the age of each boy, supposing all to be of the same age, that the sum of their ages may equal the sum of the ages in the question.

Sum of ages
$= 17 + 32 + 62 + 14\frac{3}{4} + 29 + 68\frac{3}{4} + 122\frac{1}{2} + 20 = 366$
$\therefore$ average age $= 366$ yrs. $\div 27 = 13\frac{5}{9}$ years.

Ex. LXXV.

(1) The highest temperature registered in the shade on

Monday 13th July, 1868, in the following towns, was :—Ottawa, 104; Montreal, 96; Toronto, 92; New York, 90; Buffalo, 82; New Orleans, 81. Find their average highest temperature?

(2) On Sunday I spent no money, on Mond. $4.25, on Tues. $5.75, on Wed. $6.60, on Thurs. $7.80, and Frid. $3.50, on Sat. $5.58; find my average daily expenditure during the week?

(3) The highest temperature registered in the shade in the week ending on Midsummer-day, 1865, in the following towns, was :—Birmingham, 87·8; Manchester, 87·7; London, 87·6; Bristol, 86·8; Leeds, 85·0; Salford, 84·5; Dublin, 83·8; Edinburgh, 78·0; Liverpool, 77.9; Glasgow, 77·6. Find their average highest temperature?

(4) In a school, 17 children average 6 years.; 26, 7½ yrs., 35, 9¼ yrs.; 20, 10 yrs.; and 8 12¼ yrs. Find the average age of all the children.

(5) The average age of 27 men is 57 years, that of the first eleven is 53 years, and that of the last eight 59¼ years. Find the average age of the rest.

(6) The populations of 3 towns in 1851 were 31326, 42324, and 6706; in 1861 the first two had increased 12, and 10 per cent. respectively, and the last had decreased 18 per cent.; find the average population of the 3 towns in 1861.

(7) A tradesman's average annual gain from the year 1853 to 1863, both inclusive, was £184. 11s. 6d.; in 1853 he lost £76. 8s. 4d., and in 1864 he gained £151. 9s. 10d. What was his average annual gain from 1854 to 1864, both inclusive?

DIVISION INTO PROPORTIONAL PARTS.

137. *To divide a given number into parts, which shall be proportional to certain other given numbers.*

This is an application of the **Rule of Three**; still it may be well to state a general Rule, by which such Ex^s. may be worked.

RULE. **As the sum** of the given parts : any one of them :: the entire **quantity to** be divided : the corresponding part of it.

This statement **must be** repeated for each of the parts, or at all events for **all but the last part, which** may either be

ARITHMETIC.

found by the Rule, or by subtracting the sum of the values of the other parts from the entire quantity to be divided.

Ex. 1. Divide 40 dollars among A, B, C, so that their shares may be as 7, 11, and 14 respectively.

By the Rule. Sum of shares $= 7 + 11 + 14 = 32$.
∴ $32 : 7 :: \$40 : A\text{'s sh}^\text{e}.$; $32 : 11 :: \$40 : B\text{'s sh}^\text{e}.$;
whence $A\text{'s sh}^\text{e}. = \$8.75$, $B\text{'s sh}^\text{e}. = \$13.75.$,
$C\text{'s sh}^\text{e}. = \$40 - \$(8.75 + \$13.75) = \$17.50$.

Ex. 2. Divide £45 among A, B, C, and D, so that A's share : B's share :: 1 : 2, B's : C's :: 3 : 4, and C's : D's :: 4 : 5.

The L. C. M., of 1, 2, 3, 4, and 5. is 60, ∴ if D has 60 shares, C will have $\frac{4}{5}$ of 60, or 48 ; B will have $\frac{3}{4}$ of 48, or 36 ; and A will have $\frac{1}{2}$ of 36, or 18.

∴ $(18 + 36 + 48 + 60)$, or $162 : 18 :: £45 : A\text{'s sh}^\text{e}.$;
whence $A\text{'s sh}^\text{e}. = £5$. Similarly $B\text{'s} = £10$. $C\text{'s} = £13$. 6s. 8d., and $D\text{'s} = £16$. 13s. 4d.

FELLOWSHIP OR PARTNERSHIP.

138. FELLOWSHIP or PARTNERSHIP is a method by which the respective gains or losses of partners in any mercantile transactions are determined.

Fellowship is divided into SIMPLE and COMPOUND FELLOWSHIP: in the former, the sums of money put in by the several partners continue in the business for the same time; in the latter, for different periods of time.

The Rule in the last Art. applies for SIMPLE FELLOWSHIP.

Ex. Two merchants, A and B, form a joint capital; A puts in $240, and B $360; they gain $80. How ought the gain to be divided between them?

$\$(240 + 360) : \$240 :: \$80 : A\text{'s sh}^\text{e}.$ in $'s
∴ $A\text{'s sh}^\text{e}. = \$32$, and $B\text{'s sh}^\text{e}. = \$(80 - 32) = \$48$.

COMPOUND FELLOWSHIP.

139. RULE. Reduce all the times into the same denomination, and multiply each man's stock by the time of its continuance, and then state thus :

The sum of all the products : each particular product :: the whole quantity to be divided : the corresponding share.

Ex. A and B trade together; A puts in $300 for 9 mo., and B $240 for 6 mo; they gain $115. How ought they to divide it?

By the Rule.

$(300 \times 9 + 240 \times 6) : \$(300 \times 9) :: \$115 : A\text{'s sh}^e.,$
$(300 \times 9 + 240 \times 6) : \$(240 \times 6) :: \$115 : B\text{'s sh}^e.,$
whence, $A\text{'s sh}^e. = \$75$, and $B\text{'s} = \$40$.

Reason. $300 for 9 mo. = 9 times $300 for 1 mo., and $240 for 6 mo. = 6 times $240 for 1 mo. : the **example** then becomes one of Simple Fellowship.

EQUATION OF PAYMENTS.

140. When a person owes another several sums of money, due at different times, the Rule by which we determine the just time when the whole debt may be discharged at one payment, is called the EQUATION OF PAYMENTS.

Note. It is assumed in this Rule that the sum of the interests of the several debts for their respective times equals the interest of the sum of the debts for the equated time.

RULE. Multiply each debt into the time which will elapse before it becomes due, and then divide the sum of the products by the sum of the debts; the quotient will be the equated time required.

Ex. 1. *A* owes *B* $100, whereof $40 is to be paid in 3 mo., and $60 in 5 **mo.**; find the equated **time.**

By the Rule,

$$\text{equated time in mo.} = \frac{40 \times 3 + 60 \times 5}{40 + 60} = \frac{420}{100} = 4\tfrac{1}{5}.$$

Ex. 2. *A* owed *B* $10, to be paid at the end of 9 mo.; he pays however $2 at the end of 3 mo., and $3 at the end of 8 mo.; when ought the remainder to be paid?

In this case, $2 \times 3 + 3 \times 8 + 5 \times$ no. of mo. req^d. $= 10 \times 9$, or $6 + 24 + 5 \times$ no. of mo. req^d. $= 90$;

or, $30 + 5 \times$ no. of mo. req^d. $= 90$, or $5 \times$ no. of mo. req^d. $= 90 - 30$, or 60, $\therefore$ no. of mo. req^d. $= 12$.

Ex. LXXVI.

(1) Divide (1) 1008 into 3 parts, which shall be to each other as the numbers 2, 3, 4, respectively. (2) $260 into 3 parts, which shall be to each other as 5, 11, and 16. (3) 145 ac. 3 ro. 33 po. between two persons in the ratio of 5 : 6. (4) £110 between 4 persons, whose shares shall be as $\tfrac{1}{2}$, $\tfrac{1}{3}$, $\tfrac{1}{4}$, and $\tfrac{1}{5}$.

(2) (1) *A*, *B*, and *C* contribute to a fund $320, $560, $720, respectively. How are they to divide a profit of $680? (2) *A*. who has £422. 10s., owes *B*, £175; *C*, £210; and *D*, £265; what sum ought *C* to receive?

(3) Sugar being composed of 48·856 per cent. of oxygen, 43·265 per cent. of carbon, and the rest hydrogen; how many lbs. of each of these materials are there in 1 ton of sugar?

(4) Archimedes discovered that the crown made for King Hiero consisted of gold and silver in the ratio of 2 : 1. How much per cent. was gold, and how much per cent. was silver?

(5) Find the equated time of payment of $150 due in 2 mo., $210 due in 6 mo., and $120 due in 7 mo.

(6) A owes B $1000 to be paid at the end of 6 mo.; A pays $400 at the end of 3 mo.; when ought he to pay the remainder?

(7) A, B, and C remained partners for 2 years; A brought in $4000, which remained the whole time; B began with $300, and 6 months after put in $300 more; C began with $200, and one year after put in $500 more. The whole gain was $7960. Determine each partner's share.

(8) A is a working, B a sleeping partner in a bookseller's business. Their capital amounts to £6400; of which £2400 belongs to A, the rest to B. Their profits, at the end of the first year, amounted to £1600. A receives 10 per cent. of the profits for managing the business. How ought the remaining part of the profits to be divided?

(9) A, B, and C rent a field for $60; A puts in 20 horses, B 15 oxen, and C 10 sheep; supposing the keep of a horse, ox, and sheep to be in the ratio of 3, 2, and 1; shew how the rent should be divided.

(10) Some broth was distributed among a certain number of old men, 9 widows, and 6 single women; the men had twice as much broth given among them as was given among the women; also an old man's share was to a widow's share :: 6 : 5, and a widow's share to a single woman's share :: 10 : 9. Each single woman received $1\frac{1}{2}$ pints. How many old men were there?

SQUARE ROOT.

141. The SQUARE of a given number is the product of that number multiplied by itself. Thus 6×6 or 36 is the square of 6, or $36 = 6^2$. Art. 86.

142. The SQUARE ROOT of a given number is a number,

SQUARE ROOT. **147**

which, when multiplied by itself, will produce the given number. Thus 6 is the square root of 36; for $6 \times 6 = 36$.

The square root of a number is sometimes denoted by placing the sign $\sqrt{}$ before the number, or by placing the fraction $\frac{1}{2}$ above the number a little to the right. Thus $\sqrt{36}$, or $(36)^{\frac{1}{2}}$ denotes the square root of 36; so that $\sqrt{36}$, or $(36)^{\frac{1}{2}} = 6$.

143. *Rule for extracting the Square Root of a number.*

Place a point or dot over the units' place of the given number; and thence over every second figure to the left of that place; and thence also over every second figure to the right, when the number contains decimals, annexing a cypher when the number of decimal figures is odd; thus dividing the given number into periods. The number of points over the whole numbers and decimals respectively will shew the number of whole numbers and decimals respectively in the square root.

Find the greatest number whose square is contained in the first period at the left; this is the first figure in the root, which place in the form of a quotient to the right of the given number. Subtract its square from the first period, and to the remainder bring down, on the right, the second period.

Divide the number thus formed, omitting the last figure, by twice the part of the root already obtained, and annex the result to the root and also to the divisor.

Then multiply the divisor, as it now stands, by the part of the root last obtained, and subtract the product from the number formed, as above mentioned, by the first remainder and second period.

If there be more periods to be brought down, the operation must be repeated.

Ex. 1. Find the square root of 1369.

$$\begin{array}{r|l} & 1\dot{3}6\dot{9}\,(\,37 \\ 3^2 = & 9 \\ \{2 \times 3 = 6\}\ 67 & \overline{469} \\ & 469 \end{array}$$

After pointing, according to the Rule, we take the first period, or 13, and find the greatest number whose square is contained in it. Since the square of 3 is 9, and that of 4 is 16, it is clear that 3 is the greatest number whose square is contained in 13, therefore place 3 in the form of a quotient to the right of the given number. Square this number, and put down the square under the 13; subtract it from the 13, and to the remainder 4 affix the next period 69, thus forming the number 469. Take 2×3, or 6, for a divisor; di-

K

ARITHMETIC.

vide the 469, omitting the last figure, that **is, divide the 46 by the 6, and we obtain 7.** Annex **the 7** to the **3** before obtained, and to the divisor **6 ; then multiplying** the 67 by the 7 we obtain 469, which **being subtracted from the** 469 before formed, leaves **no remainder ; therefore 37 is the** square root of 1369.

Ex. 2. Find the square root of 282475249.

```
            282475249 ( 16807
            1
{2×1=2}   26 | 182        18÷2=9, but 9 will be found too
             | 156        large, so also 8 or 7. ∴ try 6.
{2×16=32} 328 | 2647      264÷32=8
              | 2624
{2×168=336} 33607 | 235249   336 is greater than 235 ; ∴ put
                  | 235249   0 after the 8 in the quotient,
                              and the 6 in the divisor, bring
```
down the next period. Then 23524 ÷ 3360 = 7.

Ex. 3. Find **the** square root **of 7·929856.**

```
       7·929856 ( 2·816
       4
    48 | 392
       | 384
   561 | 898
       | 561
  5626 | 33756
       | 33756
```

Place **the** first dot over the 7, the units' place of whole numbers, and then **over** every second figure to the right.

There is 1 dot over the integral part, and 3 dots over the dec¹. part, ∴ the root is 2·816.

Ex. 4. Find the square **root of ·001 to 3 places of decs.**

```
           ·001000 ( ·031
           9
{2+3=6}  61 | 100
            | 61
            | 39
```

We affix 3 cyphers in order to have 3 periods, and ∴ 3 dec¹. places in root ; since there is **no** number in the **units'** place, the first dot will be **over the** second cypher from the units' **place, and since** first period **is ·00** we place ·0 as the first figure **in** the root.

Ex. 5. Find the square root of $\frac{529}{2401}$.

```
     529 ( 23            2401 ( 49
     4                   16
  43 | 129             89 | 801
     | 129                | 801   ∴ sq. root = 23/49
```

CUBE ROOT.

Ex. 6. Find the square root of $\dfrac{5}{7}$ to 3 places of dec¹ˢ.

$$\tfrac{5}{7} = \cdot 714285\ldots;\qquad \overset{\cdot}{7}14285\ (\ \cdot 845\ldots$$

$$\begin{array}{r|l}
 & 64 \\
164 & 742 \\
 & 656 \\
1685 & 8685 \\
 & 8425 \\ \hline
 & 260
\end{array}$$

$\therefore$ sq. root of $\dfrac{5}{7} = \cdot 845\ldots$

Ex. LXXVII.

Find the square roots of (1) 196; 289; 625. (2) 841; 900; 1764. (3) 2401; 7569; 9604. (4) 12321; 40000; 388129. (5) 494209; 582169; 259081. (6) 1234321; 28547649. (7) 62504836; 33016516; 49112064. (8) 182493081; 47·61. (9) ·008836; 445·336609. (10) ·000633679929; ·0000000009.

Find the square roots, each to four places of decimals, of (11) 51; ·51. (12) 5·1; ·051. (13) 806·52; 96304·993.

Find the square roots, each to 3 places of decimals where the root does not come out exactly, of (14) $\cdot \overset{\cdot}{3}$. (15) $\cdot 02\overset{\cdot}{7}$. (16) $4\tfrac{9}{9}$. (17) $\dfrac{2304}{3481}$. (18) $\dfrac{4\cdot 41}{\cdot 64}$.

(19) A father left his child a box, containing sovereigns, and shillings; the sovereigns were worth as many times the shillings, as the shillings were worth the box; the value of the box was 2s. 6d., and there were 5832 sovereigns in the box. How many shillings were there?

CUBE ROOT.

144. The CUBE of a given number is the product which arises from multiplying that number by itself, and then multiplying the result again by the same number. Thus $6 \times 6 \times 6$, or 216, is the cube of 6; or $216 = 6^3$. Art. 86.

145. The CUBE ROOT of a given number is a number, which, when multiplied into itself, and the result again multiplied by it, will produce the given number. Thus 6 is the cube root of 216; for $6 \times 6 = 36$, and $36 \times 6 = 216$.

The cube root of a number is sometimes denoted by plac-

150   ARITHMETIC.

ing the sign $\sqrt{}$ before the number, or placing the fraction $\frac{1}{3}$ above the number, a little to the right. Thus $\sqrt[3]{216}$ or $(216)^{\frac{1}{3}}$ denotes the cube root of 216 ; so that $\sqrt[3]{216}$ or $(216)^{\frac{1}{3}} = 6$.

146. *Rule for extracting the Cube Root of a number.*

Place a point or dot over the units' place of the given number, and thence over every third figure to the left of that place ; and thence also over every third figure to the right, when the number contains decimals, affixing one or two cyphers, when necessary, to make the number of decimal places a multiple of 3 ; thus dividing the given number into periods. The number of points over the whole numbers and decimals respectively will shew the number of whole numbers and decimals respectively in the cube root.

Find the greatest number whose cube is contained in the first period at the left; this is the first figure in the root, which place in the form of a quotient to the right of the given number.

Subtract its cube from the first period, and to the remainder bring down, on the right, the second period.

Divide the number thus formed, omitting the two last figures, by 3 times the square of the part of the root already obtained, and affix the result to the root.

Now calculate the value of 3 times the square of the first figure in the root (which of course has the value of so many tens) + 3 times the product of the two figures in the root + the square of the last figure in the root. Multiply the value thus found by the second figure in the root, and subtract the result from the number formed, as above mentioned, by the first remainder and the second period. If there be more periods to be brought down the operation must be repeated.

Ex. 1. Find the cube root of 15625.

$$15625(25$$

$$2^3 = 8$$

$$3 \times 2^2 = 12$$

$$3 \times (20)^2 = 3 \times 400 = 1200$$
$$3 \times 20 \times 5 = 300$$
$$5^2 = 25$$
$$\overline{1525}$$
$$\text{Multiply by } 5$$
$$\overline{7625}$$

$\overline{7625}$

$\overline{7625}$

After pointing we take the first period, or 15, and find the greatest number whose cube is contained in it. Since the cube of 2 is 8, and that of 3 is 27, it is clear that 2 is the greatest number whose cube is con-

CUBE ROOT. 151

tained in 15 ; ∴ place **2 in the form of a quotient to the right of the** given number.

Cube 2, and put down its cube, viz. 8, under the 15; subtract it from the 15, and to the rem^r. 7 affix the next period 625, thus forming the number 7625. Take 3×2^2, or 12, for a divisor; divide 76 by 12, 12 is contained 6 times in 76; but when the other terms of the divisor are brought down 6 would be found too great, therefore try 5. Affix the 5 to the 2 before obtained; and calculate the value of $3 \times (20)^2 + 3 \times 20 \times 5 + 5^2$, which is 1525; multiplying 1525 by 5 we obtain 7625, which being subtracted from 7625 before formed leaves no rem^r.; ∴ 25 is the cube root req^d.

Ex. 2. Find the cube root of 219·365327791.
Place the first dot over the 9 in the units' place.

$$219\cdot365327791\ (\ 6\cdot031$$

$$
\begin{array}{rr}
6^3 = & 216 \\
3 \times 6^2 = & 108 \\
3 \times (60)^2 = & 10800 \\
3 \times (600)^2 = & 1080000 \\
3 \times 600 \times 3 = & 5400 \\
3^2 = & 9 \\
\hline
& 1085409 \\
& 3 \\
\hline
& 3256227
\end{array}
$$

3365
3365327   33 is not divisible by 108;
bring down the next period and affix 0 to the root; the trial divisor will then be $3 \times (60)^2 = 10800$, and $33653 \div 10800$ goes 3 times, try 3.

3256227

$$
\begin{array}{rr}
3 \times (603)^2 = & 1090827 \\
3 \times (6030)^2 = & 109082700 \\
3 \times 6030 \times 1 = & 18090 \\
1^2 = & 1 \\
\hline
& 109100791
\end{array}
$$

109100
109100791   bring down next period $1091007 \div 1090827$ goes once, try 1.

109100791

∴ 6·031 is the cube root required.

Ex. 3. Find the cube root of ·000007 to three places of decimals.

$$\cdot 000007000\ (\ \cdot 019$$

$$
\begin{array}{rr}
& 1 \\
3 \times 1^2 = 3 & 6000 \\
3 \times (10)^2 = & 300 \\
3 \times 10 \times 9 = & 270 \\
9^2 = & 81 \\
\hline
& 651 \\
& 9 \\
\hline
& 5859
\end{array}
$$

5859
141

152   *ARITHMETIC.*

147. Higher **roots than the square** and cube can sometimes be extracted **by means of the** Rules for square and cube root; **thus the 4th root is found** by taking the square root of the square root; the 6th root by taking **the** square root of the cube root, and so on.

Ex. LXXVIII.

Find the cube roots of
    (1)   1728 ; 8000 ; 5832.
    (2)   74088 ; 421875 ; 778688.
    (3)   912673 ; 1092727.
    (4)   134217728 ; 64·481201.
    (5)   444194·947 ; ·000202262003.
    (6)   131·019108039 ; 4085184S8000.

Find the cube roots, to three places of decimals in those cases where the root does not terminate, of

    (7) $\frac{27}{64}$.   (8) $\frac{1}{8}$.   (9) $3\frac{3}{8}$.   (10) 1.
    (11) ·1.   (12) ·01.   (13) 10.   (14) ·037.

MISCELLANEOUS QUESTIONS.

Ex. LXXIX.

PAPER I.

1. Subtract 2057312 from 5287201, **and** 2057312 again from the remainder. Explain how this is the same as dividing 5287201 by 2057312.

2. (1) Reduce 553553 oz. to tons, cwts., &c. (cwt. = 112 lbs.)   (2) Find the proportion of the Avoird. and Troy oz., when the respective lbs. are as 175 : 144.

3. Find, by Practice, the cost of 16 cwt., 3 qrs., 16 lbs. at £2. 7 cents a cwt., (112 lbs. = cwt.) £1 being = 10 florins = 100 cents = 1000 mils.

4. Define (1) the G. C. M., (2) the L. C. M., of two or more numbers, (3) a Vulgar Fraction. Find the G. C. M. of 20803 and 67273 ; and the L. C. M. of 8, 9, 10, 12, 15, 18, 35 and 84.

5. (1) Add together $\frac{3}{8}$ of $\frac{1}{7}$ of $99\frac{1}{4}$, $\frac{2}{5}$ of $\frac{3}{8}$ of $69\frac{3}{15}$, $\frac{7}{9}$ of $\frac{2}{3}$ of $306\frac{1}{4}$.   (2) Express 13s. $1\frac{1}{2}d.$ as the fraction of $\frac{3}{4}$ of $1\frac{1}{2}$ guinea.   (3) Find the value of $\frac{107}{448}$ ton (cwt. = 112 lbs.).

MISCELLANEOUS. 153

6. State the Rule for the division of one decimal by another. Divide (1) 7792·2 by ·37, (2) ·0077922 by 370; verify each result by vulgar fractions.

PAPER II.

1. Define Interest, Simple and Compound. How does Interest differ from Discount? Find (1) the int. on $7300 at $3\frac{3}{4}$ per cent. for 120 days, (2) the discount on £3204. 14s. 1d. at $3\frac{1}{2}$ per cent. simp. int. for $2\frac{2}{3}$ yrs.

2. A house built for $2656 is sold for $3320, find the gain per cent. If it had been built for $3320 and sold for $2656, find the loss per cent.? Why do the rates differ?

3. Define a square. Find (1) the sq. root of 930372004, (2) the cub. root of 16777216, (3) the perimeter of a square whose surface is 2533 sq. ft., ·64 sq. in.

4. Multiply 365 separately by 5, by 20, and by 300, and add the products together. Point out how the ordinary method of multiplying 365 by 325 agrees step by step with the above.

5. Define prime and composite numbers. Resolve 22932 into its prime factors.

6. A person left Toronto for Guelph at 9 A.M., and travelled the first 20 miles by rail, at the rate of $22\frac{1}{2}$ miles an hour; he then walked the remaining 32 miles at $\frac{1}{3}$ of that rate. At what o'clock did he arrive?

PAPER III.

1. A and B fire at targets, having 55 cartridges each. A fires twice in 3 minutes, and B three times in 5 minutes; how many times will B have to fire after A has finished?

2. (1) Convert $\dfrac{17}{20 \times 8}$ into a decimal; why is the result a terminating, and not a recurring decimal? (2) Express 3s. $0\frac{1}{2}d$. as the decimal of £5. (3) Which is greater, ·36 of a guinea, or ·36 of £1? (4) By how much?

3. What sum of money will amount to $552.50 in 15 mo. at 5 per cent. simp. int.?

4. A room whose height is 11 ft., and length twice its breadth, takes 143 yds. of paper 2 ft. wide for its four walls; how much carpet will it require?

5. Two clocks strike 9 together on Tuesday morning.

On Wednesday morning one wants 10 minutes to 11 when the other strikes 11. How much must the slower be put on that they may strike 9 together in the evening?

6. A person bought 43 shares in a coal mine at 35¼, and and kept them till they declined to 11¹, when he sold out and bought with the proceeds 6 per cent. bank stock at 28 premium; find his annual income from the latter investment.

PAPER IV.

1. Define a fraction, and shew from your definition that $\frac{1}{2} = \frac{3}{6}$ (1) Add together $\frac{1}{8}$, $\frac{2}{8}$, $\frac{5}{16}$, and $\frac{3}{7}$; and find what fraction the sum is of $1\frac{2}{3}$ of $\frac{4}{2\frac{2}{3}}$ (2) How many times can ·027 be taken from 3·33? What fraction is the remainder of the former?

2. A person left a sum of money which was divided equally amongst 43 poor people, such that, after a deduction of 6d. in the pound, each received £3. 3s. 4½d. What sum did he leave?

3. (1) If the carriage of 13 cwt., 2 qrs, 19 lbs. for 35 miles cost £4. 17s. 6d., what must be paid for the conveyance of 41 cwt., 1 lb. for 49 miles? (A cwt = 112 lbs.) (2) A bankrupt owes $2085, of which $235 is due to A, $325 to B, $525 to C, and the rest to D. How much must he pay in the $ so that D may receive as much as is due to C?

4. A merchant buys 2 butts of wine, one for £120, and one for £110, he also buys a third, and after mixing the three, retails the wine at 45s per dozen, making 12½ per cent. on his outlay. supposing the number of dozens in a butt to be 52, find the price of the third butt.

5. The price of 2 turkeys and 9 fowls is £2. 18s. 6d. and the price of 5 turkeys and 2 fowls is £4. 8s. 2d.; find the price of a turkey and a fowl.

6. How long will it take to walk round a square field containing 13 ac., 81 yds. at the rate of 3½ miles an hour?

PAPER V.

1. Find the product of the following numbers:—
(1) 3916 × 769.   (2) 98367 × 9876.   (3) 60706 × 7095.
(4) 968175 × 39078.   (5) 9487918 × 7982.

MISCELLANEOUS.

2. A merchant bought 974 yds. cloth, and sold it all for $847.38, gaining $301.94; what was the cost per yard?

3. A and B own together 120 acres, A having 24 acres more than B. A sells his share for $84 per acre. B sells his share for the same amount as A; how much does B get per acre?

4. If potatoes be bought at $20.35 and sold at $21.32 per load, how much will be made on a sale amounting to $6332.04?

5. A merchant sold 45980 bushels of grain that cost him 98 cents at a gain of 29 cents per bushel, and with the money bought 2299 head of cattle; how much did he pay for each?

6. If a milkman use a false measure containing ·93 of a pint instead of a pint, out of how much will he have cheated his customers when he has really sold 23 gallons 2 pints?

PAPER VI.

1. Find the length of a street in which the wheel of a barrow revolves exactly 150 times, the diameter of the wheel being $1\frac{1}{2}$ ft., and the ratio of the circumference to the diameter, 3·14159:1.

2. France is 128 millions of English acres, and the Pyrenees spread over it would cover it to the depth of 115 feet; find the bulk of the Pyrenees in cubic feet.

3. What is the height of a closet 8 ft. 4 in., by 6 ft. 8 in., which will exactly contain 12 boxes 4 ft. 6 in. long, 3 ft. 4 in. wide, 2 ft. 6 in. deep?

4. What sum of money must be left, in order that after a reduction of ten per cent. has been made, the remainder being invested in the 5 per cents. at $91\frac{1}{8}$, may give a yearly income of $100?

5. A ship worth $6000 is entirely wrecked. $3000 belonged to A, $2000 to B, and the rest to C. What are the respective losses to A, B and C, supposing the ship to have been insured only to the amount of $4500.

6. A can do a piece of work in 27 days, and B in 15 days.

A works at it alone for 12 days, *B* then works 5 days, and afterwards *C* finishes it in 4 days. Find the time in which *C* alone could do the whole work.

PAPER VII.

1. Find the product of the following numbers :—(1) 78398670 × 90785. (2) 9703978 × 679458. (3) 96870 × 708963. (4) 897463287 × 30974. (5) 906870690 × 90087.

2. Two boys go fishing : one catches 40 chub, 30 perch, and 20 trout ; the other catches an equal number of each, in all 90 fish. They sell them, a hub for 5c., a perch 8c., and a trout, 12c. ; how much does each receive ?

3. A case of strawberries contains 54 boxes, each 1 lb. in weight at 7c a box. What will be the cost of canning 2 cases, allowing 1 lb sugar at 10c to every 2 lbs. berries ?

4. Each man in an army of 60000 men gets two pairs of socks per year. How many sheep, each fleece 6 lbs., are necessary to supply wool for the socks, 1 lb. wool making 8 socks ?

5. Jones and Smith are farmers. Jones sold last year 200 bush. oats at 38c., 73 bush. peas at 81c., 580 bush. wheat at 98c, 156 bush potatoes at 29c, 138 bush. barley at 87c Smith sold 45 sheep at $5, 60 lambs at $3 30, 18 young cattle at $15 18 large cattle at $29, and 26 tons hay at $19. What sum did each receive ?

6. A merchant sold a cargo of wheat valued at $40000 for ¼ less than this amount, thus making a profit of only ⅙ on cost. At what advance on cost was the wheat valued at in the first instance ?

PAPER VIII.

1 Find the product of the following numbers :—(1) 987798640 × 10970. (2) 793289765 × 40097. (3) 7963 × 8679. (4) 874598 × 39076.

2. A shopkeeper bought $9.60 worth of steel pens at 32 cents per box, each containing 12 dozen and retailed them at 5 cents per dozen. How much did he gain on his outlay ?

MISCELLANEOUS. 157

3. A person distributes $22.68 amongst six men, eight women and twelve boys. Each woman had three times as much as each boy, and each man half as much again as each woman. Find what each received.

4. Goods were bought for 8648 dollars; there was further paid for packing, 20 dollars; for lake carriage, 55 dollars; for land carriage, 115 dollars; and for other charges, 350 dollars. The goods were then sold for 10000 dollars. What was the profit made on the sale?

5. Divide 1120 cents between three boys, Alfred, Benjamin and Charles, so that Alfred may have three times as much as Benjamin, and Charles as much as Alfred and Benjamin together.

6. In 1871 the population of England and Wales was 22704108; of Scotland, 3358613; of Ireland, 5402759; of islands in the British seas, 144430; and of the army and navy, &c., 207198. Find the total population of the United Kingdom at that date.

PAPER IX.

1. Divide (1) 6022808 by 769; (2) 1942944984 by 9876; (3) 55596055703076 by 15487; (4) 326789039400120 by 90087.

2. If a locomotive travelled from Toronto to Whitby at a uniform rate of 830 yards a minute, it could perform exactly the distance in 60 minutes; find the distance between the two places in yards.

3. Three men, A, B and C, start on a journey, each with 126 dollars in his pocket, and agree to divide their expenses equally. On their return home, A has 106 dollars, B has 56 dollars, and C has 66 dollars. What ought A to pay B and C to settle their accounts?

4. A farmer bought two farms, each of 130 acres, for 19500 dollars. What is the value of an acre of each farm, if two acres of one be worth three acres of the other?

5. A gentleman in Toronto remits $10696.93½ to a friend in London. How much does it amount to in London, exchange at 109½, commission ⅔ % extra?

6. Brown, in London, has £715 stg. He sends it to a friend in Toronto. How much does the friend realize, exchange at 109½, commission ½ % extra?

PAPER X. (ADMISSION TO HIGH SCHOOLS.) 1877.

1. How often is 6 yds. 2 ft. contained in 25 furlongs?

2. If I buy 3 bushels, paying 5 cents for every 3 quarts, and sell at a profit of   cents per gallon, find the selling price of the whole.

3. Simplify $\dfrac{2\frac{1}{3}+\frac{5}{8} \text{ of } 12-\frac{5}{6}}{3\frac{1}{4}\times \cdot 01+1\frac{9}{10}} \times \dfrac{11}{3\frac{2}{3}} \times \dfrac{18\frac{1}{2}+5\frac{8}{15}-22\frac{3}{30}}{\dfrac{1}{1\frac{1}{27}}\div(2\frac{7}{16}-\frac{5}{8}+4)}$

4. Reduce 2 hrs. 20 min. to the decimal of 3⅓ weeks.

5. A sum of money was divided among A, B and C. A received ⅔ of the sum; B, $20 less than ⅝ of what was left; and the remainder, which was ¾ of A's share, was given to C. Find the sum divided.

6. Trees are planted 12 feet apart around the sides of a rectangular field (40 rods long) containing two acres. Find the number of trees.

7. I buy a farm containing 80 acres, and sell ¾ of it for ⅔ of the cost of the farm; I then sell the remainder at $60 per acre, and neither gain nor lose by the whole transaction. Find the cost of the farm.

8. Find the amount of the following bill of goods:—18¾ cords of wood, at $3.50 per cord. 16 yards of cloth, at $1.12½ per yard. 12 bush. 25 lbs. of wheat, at $1.20 per bush. 1,400 feet of lumber, at $12.50 per thousand. 65 tons 12 cwt. of coal, at 30 cents per cwt.

PAPER XI. (ADMISSION TO HIGH SCHOOLS.) 1878.

1. Define prime number, multiple of a number, highest common factor of two or more numbers, ratio between numbers. Find the prime factors of 1260.

2. The quotient is equal to six times the divisor; the divisor is equal to six times the remainder, and the three together, plus 45, amount to 561. Find the dividend.

MISCELLANEOUS.

3. I sell $12\frac{1}{2}$ tons of coal for $80, which is one-seventh more than the cost. Find the gain per cwt.

4. $\cdot 001 \times \cdot 001 \div \cdot 0001$.

5. A cistern is two-thirds full; one pipe runs out and two run in. The first pipe can empty it in eight hours, the second can fill it in twelve hours, and the third can fill it in sixteen hours. There is also a leak half as large as the second pipe. In how many hours will the cistern be half full?

6. Ten men can do a piece of work in twelve days. After they have worked four days, three boys join them in the work, by which means the whole is done in ten days. What part of the work is done by one boy in one day?

7. I buy a number of boxes of oranges for $600, of which 12 boxes are unsaleable. I sell two-thirds of the remainder for $400, and gain on them $40. How many boxes did I buy?

8. Find the total cost of the following:—Cutting a pile of wood 80 ft. long, 6 ft. high, and 4 ft. wide, at 60c. per cord. Digging a cellar 44 ft. long, 30 ft. wide, and 8 ft. deep, at 18c. per cubic yard. Plastering a room 24 ft. long, 16 ft. wide and 10 ft. high, at 15c. per square yard. Sawing 6800 shingles, at 40c. per 1000.

The Independent Method, or the Method of Reduction to the Unit, introduced at page 89, may with advantage be employed to solve questions which can also readily be done by the Rule of Three. We subjoin a few more examples, showing how to apply the method referred to.

1. If 27 men build a house in 63 days, in how many days will 42 men do the same?

$$\quad 27 \text{ men build a house in } 63 \text{ days};$$
$$\therefore \quad 1 \text{ man } \quad \text{``} \quad \text{``} \quad 63 \times 27 \text{ days};$$
$$\therefore \quad 42 \text{ men } \quad \text{``} \quad \text{``} \quad \frac{63 \times 27}{42} \text{ days};$$
$$\therefore \text{ Number of days required} = \frac{63 \times 27}{42} = 40\frac{1}{2}.$$

160 ARITHMETIC.

2. A person rows down a stream in 20 minutes, but without the aid of the stream it would have taken him half an hour. What is the rate of the stream per hour, and how long would it take him to row against it?

1st. Moving with stream:
   In 20′, distance rowed $= 1\frac{1}{2}$ miles;
   ∴ in 1′, " $= \frac{3}{40}$ miles.

2nd. Moving in still water:
   In 30′, distance rowed $= 1\frac{1}{2}$ miles;
   ∴ in 1′, " $= \frac{3}{60}$ miles;
   ∴ rate of stream $= \frac{3}{40} - \frac{1}{20} = \frac{1}{40}$ miles;
   ∴ rate of stream per hour, $\frac{1}{40} \times 60 = 1\frac{1}{2}$ miles.
   Rate of stream in $1' = \frac{1}{40}$ miles,
   in still water, distance rowed $= \frac{1}{20}$ miles;
   ∴ distance rowed against stream $= (\frac{1}{20} - \frac{1}{40})$ miles
   $= \frac{1}{40}$ miles;
   ∴ time required to row $1\frac{1}{2}$ miles $= \frac{3}{2} \div \frac{1}{40} = \frac{3 \times 40}{2} =$
60′ = 1 hour.

3. At what time between 1 and 2 are the hands of a clock opposite to each other?

Let OC be the position of the hr. hand.

Let OD be the position of the min. hand.

At 1 o'clock OC overlapped OB, and OD overlapped OA.

Then BC space passed over by hr. hand, and AD space passed over by min. hand.

12 times $BC = AD$ (1).

But $AD = AB + BC + CD$.
$= 5$ min. $+ BC + 30$ min.

MISCELLANEOUS.

∴ substituting this value of AD for AD in (1), we have
12 times $BC = 35$ min. $+ B.C.$
∴ 11 times $BC = 35$ min.,
or $BC = 35$ min. $\div 11 = 3\frac{2}{11}$ min.
∴ $AD = 35$ min. $+ 3\frac{2}{11}$ min.
∴ time required is $38\frac{2}{11}$ min. past 1 o'clock.

In connection with the above we give the following statement: Since the minute hand moves twelve times as fast as the hour hand, therefore in 12 minutes the minute hand gains 11 minute spaces on the hour hand.

4. *The hands of a clock are together at 12, when will they be together again?*

The time must be after one; therefore the minute hand has 5' to gain.

11 minute spaces gained in 12';
∴ 1 minute space gained in $1\frac{2}{11}'$;
∴ 5 minute spaces gained in $\dfrac{12 \times 5'}{11}$;
∴ time required is $5\frac{5}{11}'$ past 1.

5. *After paying an income tax of $10 on a $100, a person has $2700 a year. What was his entire income?*

10 on a 100 = $\frac{1}{10}$ on a unit;
∴ $\frac{9}{10}$ of every unit of income left;
∴ $\frac{9}{10} = \$2700$;
∴ $\frac{1}{10} = \$300$;
∴ 1, or whole income $= \$300 \times 10 = \$3000$.

6. *A stock of provisions will serve 75 men for 30 days. How many men must leave in order that the stock may hold out 45 days for those left?*

Provisions last 30 days for 75 men;
∴ " " 1 day for 75×30 men;
∴ " " 45 days for $\dfrac{75 \times 30}{45}$ men, or 50 men.

Hence the number of men who must leave $= 75 - 50 = 25$.

Exercise LVI., &c., furnish examples.

ARITHMETIC.

EXCHANGE is the Rule by which we find how much money of one country is equivalent to a given sum of another country, according to a given Course of Exchange.

By the COURSE OF EXCHANGE is meant the *variable* sum of the money of any place which is given in exchange for a *fixed* sum of money of another place.

By the PAR OF EXCHANGE is meant the intrinsic value of the coin of one country as compared with a given fixed sum of money of another.

ARBITRATION, or COMPARISON OF EXCHANGES, is the method of fixing upon the rate of exchange, called the PAR OF ARBITRATION, between the first and last of a given number of places, where the course of exchange between the first and second, second and third, &c., of these places is known. It is called SIMPLE or COMPOUND ARBITRATION, as three or more places are concerned. (For fuller information on Exchange, see *Advanced Arithmetic*, p 227, &c.)

By an Act of Parliament passed many years ago, the sovereign was declared to be only equal in value to $4.44, or £9 (sterling) = $40; and this is the value which is almost invariably quoted in mercantile transactions; the premium on this depreciated value of the sovereign which will make it equal to its intrinsic value, is $9\frac{1}{2}$ per cent.

1. A person has to transmit to Britain £450 stg.; the rate of exchange is at 6 per cent. premium, and he is charged $\frac{1}{2}$ per cent. for commission. What will the bill of exchange cost him in our currency?

By old statute, £9 = $40;

$$\therefore £1 = \$\tfrac{40}{9}.$$

Rate of exchange to the buyer is $106 + \tfrac{1}{2} = 106\tfrac{1}{2}$;

$$£1 = \$\tfrac{40}{9} \times \frac{106\tfrac{1}{2}}{100};$$

$$£450 = \$\tfrac{40}{9} \times \frac{213}{200} \times 450,$$

$$= \$2130.$$

Hence the bill of exchange costs the buyer $2130.

MISCELLANEOUS. 163

2. A person sold in Paris a bill worth in London £1275 15s. for 32148f. 90c. What was the course of exchange between these two cities.

$$£1275\cdot75 = 32148\cdot 90f. ;$$
$$\therefore £1 = \tfrac{32148\cdot90}{1275\cdot75}f.$$
$$= 25f.\ 20c. :$$

therefore the course of exchange is 25f. 20c. for £1 stg.

PAPER XII.

1. What would a currency draft on New York for $504 cost in gold, if it be purchased when gold is quoted at ⅝ % premium, the broker charging ¼ % commission?

2. How much gold would one get for $1284, U. S. currency selling at 2 % discount?

3. What would a person have to pay for $400 U. S. currency at 99?

4. You sell $1127 in gold for currency—gold = 102. How much do you receive?

5. A, in Toronto, owes B, in London, £360 stg. Exchange, 110¼. What will be the cost of a draft to cover the amount?

6. Jones, in London, sells 60 shares, £100 each, M. R'y stock, at a premium of 9 %, and invests the proceeds in O. B. stock, in Toronto, at 96. Exchange between London and Toronto, 109. How many dollars does he invest in O. B. stock.

7. A Toronto house owes £278 18s. 9d. in Manchester; how much will be required to discharge the debt in Canadian currency, rate of exchange being at 8¾ % premium?

8. A merchant in Montreal has to pay a bill of 1387f. 18c. in Paris. Find the amount he will have to remit for payment of the bill, it being known that the sovereign is worth 25f. 20c., and exchange on England in Montreal at a premium of 7⅞ per cent.

9. If 88800 are required in Toronto to pay £1800 in London, England, find the rate of exchange between the two cities.

10. A traveller for Paris wishing to provide himself with French money, calls at a broker and is informed that the sovereign in London is worth 25f. 25c., rate of exchange on London, 8½ premium, and ½ per cent. commission. Find the sum in French money he ought to receive for $500 of our money.

PAPER XIII.

1. Divide
    (1) 366170794144410 by 160388.
    (2) 757325614476 by 9487918.

2. If a gallon contain 277·274 cubic inches, and a cubic foot of water weigh 1000 ounces, what quantity in gallons and what weight of water in pounds will fill a rectangular cistern 5 feet long, 3½ feet wide, and 2 feet 9 inches deep?

3. Find the depth of the circular cistern which would hold the same quantity of water as that in question 2, supposing the diameter to be 6 feet.

4. A cubical box exactly holds 64 shot, each 3 inches in diameter. Find how many cubic inches are empty in the box when it is full of shot.

5. Find the length of the side of a square whose area is equal to that of a rectangle, the sides of which are 94·28 and 6720 yards.

6. Add together one million one thousand and ten, fifty thousand five hundred and five, ten millions, five hundred thousand and fifty, seventy millions seven hundred thousand and seventy, eight billions eight hundred thousand and eight. From the sum take 51643, and divide the remainder by 808.

PAPER XIV.

1. Divide
    (1) 3694875009838660 by 1978.
    (2) 9768757832914415 by 389.

MISCELLANEOUS.

2. Write Avoirdupois weight.
Express 1 cwt. 2 qrs. 15 lbs., in Troy Weight.

3. In 2784583 inches how many miles, furlongs, poles, &c., are there? Find how long a man going at the rate of 4 miles an hour would take to walk the number of miles, &c., in your result?

4. Define the least common multiple and the greatest common measure of two numbers. If the greatest common measure of two numbers be 103, and their least common multiple be 14729, find the numbers.

5. Define a vulgar fraction. Distinguish between a vulgar and a decimal fraction. Multiply together, expressing the resulting fraction in its lowest terms, $1\frac{3}{11}$, $1\frac{9}{14}$, $\frac{4}{7}$. $9\frac{5}{12}$, and $\frac{6}{113}$.

6. Divide ·238095 by ·3428571, and extract the square root of the quotient.

PAPER XV

1. Write square or land measure. How many square inches in 2 ac. 3 r. 5 p. 5 yds.?

2. Define simple interest and amount. At what rate per cent. per annum will a sum of money double itself at simple interest in 10 years?

3. A school section is valued at $13740. The section is required to raise by rate a sum of $820.40. What is the rate per $1?

4. A person invests $6477 in the 6 per cent. Dominion of Canada stock at $101\frac{1}{4}$, and when it has risen to 106 he sells out and invests the proceeds in a $4\frac{1}{4}$ per cent. stock at 70. Find gain or loss in income.

5. A bill on London for £960 stg. costs $4640. What is the rate of exchange?

6. What will a bill on London for £1620 cost in Toronto, exchange at $109\frac{1}{2}$, commission $\frac{1}{2}$ % extra?

The two following **Tables** may be added to those given on pp. 44–50.

TABLE OF APOTHECARIES' WEIGHT.

Table as given in *British Pharmacopœia*.

```
437½ Grains .................... make 1 Ounce.
 16  Ounces ....................      1 Pound.
```

The grain is the same as the grain Troy; the ounce is the same as the ounce Avoirdupois.

APOTHECARIES' FLUID MEASURE.

In this measure, founded on the fact that a pint of pure water weighs 20 ounces:

```
60 Minims ...... make 1 Fluid Drachm .. fl. dr.
 8 Fluid Drachms ....  1 Fluid Ounce ... fl. oz.
20 Fluid Ounces ......  1 Pint .......... o; octavius.
 8 Pints ............   1 Gallon ........ c; congius.
```

SECTION VI.

MENTAL ARITHMETIC.

48. The following *table* will be found useful.

Multiplication and Division Table.

×	1	2	3	4	5	6	7	8	9	10	11	12	13	14	15	16	17	18	19	20
1	1	2	3	4	5	6	7	8	9	10	11	12	13	14	15	16	17	18	19	20
2	2	4	6	8	10	12	14	16	18	20	22	24	26	28	30	32	34	36	38	40
3	3	6	9	12	15	18	21	24	27	30	33	36	39	42	45	48	51	54	57	60
4	4	8	12	16	20	24	28	32	36	40	44	48	52	56	60	64	68	72	76	80
5	5	10	15	20	25	30	35	40	45	50	55	60	65	70	75	80	85	90	95	100
6	6	12	18	24	30	36	42	48	54	60	66	72	78	84	90	96	102	108	114	120
7	7	14	21	28	35	42	49	56	63	70	77	84	91	98	105	112	119	126	133	140
8	8	16	24	32	40	48	56	64	72	80	88	96	104	112	120	128	136	144	152	160
9	9	18	27	36	45	54	63	72	81	90	99	108	117	126	135	144	153	162	171	180
10	10	20	30	40	50	60	70	80	90	100	110	120	130	140	150	160	170	180	190	200
11	11	22	33	44	55	66	77	88	99	110	121	132	143	154	165	176	187	198	209	220
12	12	24	36	48	60	72	84	96	108	120	132	144	156	168	180	192	204	216	228	240
13	13	26	39	52	65	78	91	104	117	130	143	156	169	182	195	208	221	234	247	260
14	14	28	42	56	70	84	98	112	126	140	154	168	182	196	210	224	238	252	266	280
15	15	30	45	60	75	90	105	120	135	150	165	180	195	210	225	240	255	270	285	300
16	16	32	48	64	80	96	112	128	144	160	176	192	208	224	240	256	272	288	304	320
17	17	34	51	68	85	102	119	136	153	170	187	204	221	238	255	272	289	306	323	340
18	18	36	54	72	90	108	126	144	162	180	198	216	234	252	270	288	306	324	342	360
19	19	38	57	76	95	114	133	152	171	190	209	228	247	266	285	304	323	342	361	380
20	20	40	60	80	100	120	140	160	180	200	220	240	260	280	300	320	340	360	380	400

149. Such questions as $7+8+3$, &c., are how many? and 29 less 7, less 6, &c., are how many? or questions in which addition and subtraction are combined, we omit; because, any teacher, by a little practice, can very easily give such exercises to the class, and, moreover, every practical teacher knows that much of the *value* of this part of the Arithmetic depends on the pupil not having seen the questions before the lesson begins.

150. *To find the value of 12 things, the value of one thing being given.*

RULE. Reckon each penny in the given value as a shilling, and each farthing as 3d.

Ex. Find the value of 12 things at $15\frac{3}{4}d.$ each.

By the Rule,

The value req^d. $= 1s. \times 15 + 3d. \times 3 = 15s.\ 9d.$

Reason for the Process.

12 things at 1d. each $= 1s.$; $\therefore$ 12 at 15d. each $= 1s. \times 15 = 15s$
12 .... $\frac{1}{4}d.$ .... $= 3d.$; $\therefore$ 12 at $\frac{3}{4}d.$ .... $= 3d. \times 3 = 9d.$;
$\therefore$ 12 things at $15\frac{3}{4}d.$ each $= 15s.\ 9d.$

151. *To find the value of 24 things, the value of one thing being given.*

RULE. Reckon each penny in the given value as 2s., and each farthing as 6d.

152. *To find the value of 48 things, the value of one thing being given.*

RULE. Reduce the given value into farthings, the result reckoned as so many shillings will be the value required.

Ex. Find the value of 48 things at $18\frac{3}{4}d.$ each.

By the Rule, since $18\frac{3}{4}d. = 75q.,$
 the value req^d. $= 75s. = £3.\ 15s.$

Reason for the Process.

 48 things at $\frac{1}{4}d. = 48q. = 1s.$;
$\therefore$ 48 things at $75q. = 1s. \times 75 = 75s. = £3.\ 15s.$

153. *To find the value of 144 things, the value of one thing being given.*

RULE. (1) Find the value of 12 things by Rule 150; then consider this value as the value of one thing, and apply Rule 150 a second time.

Ex. Find the value of 144 things at $13\frac{1}{4}d.$ each.

Value of 12 things = 13s. + 6d. = 13s. 6d.
Value of 144 things = 13s. × 12 + 6s. = 156s. + 6s. = £8. 2s.

154. The following general Rule may be given "*for finding the value of any number of things, the value of one thing being given.*"

RULE. Reckon how many dozens are contained in the given number, and how many single things remain over. Then by Rule 150, find the value of one dozen, which value multiply by the number of dozens, and add to the result the price of the single things which remained over.

Ex. Find the value of 38 things at 4s. 7d. each.
$$38 = 3 \times 12 + 2,$$
value of 12 things = £2. 8s. + 7s. = £2. 15s.
∴ ...... 12 × 3 ..... = £2. 15s. × 3 = £8. 5s.
∴ ...... 2 ..... = 4s. 7d. × 2 = 9s. 2d.
∴ ...... 38 ..... = £8. 5s. + 9s. 2d. = £8. 14s. 2d.

Ex. LXXX.

1. Find the value of 12 articles at the following prices for a single article. (1) ¾d. (2) 2d. (3) 5d. (4) 7d. (5) 11d. (6) 1½d. (7) 2¼d. (8) 3⅞d. (9) 6½d. (10) 8¼d. (11) 10½d. (12) 1s. 0¾d. (13) 1s. 4d. (14) 1s. 6¼d. (15) 1s. 9¾d. (16) 1s. 8d. (17) 1s. 11½d. (18) 1s. 2¾d. (19) 2s. 7d. (20) 3s. 0¼d. (21) 4s. 4d. (22) 6s. 1¾d. (23) 7s. 9d. (24) 8s. 5½d. (25) 11s. 7¾d. (26) 13s. 2d. (27) 16s. 3¼d. (28) 18s. 1¼d. (29) 19s. 9d. (30) 19s. 6¾d.

2. At the prices named as the value of a single article in (1) to (12) inclusive in the last question find the value of 24 articles; at the prices named in (13) to (20) inclusive find the value of 48 articles; and at the prices named in (21) to (30) inclusive find the value of 144 articles.

3. At the prices named as the value of one article in questⁿ. 1. (6) to (20) inclusive, find the value of (1) 13; (2) 21; (3) 28; (4) 35; (5) 41; (6) 44; (7) 57; (8) 72; (9) 153; (10) 182 articles.

155. *To find the value of 20 things, the value of one thing being given.*

RULE. Reckon each shilling in the given value as £1, and if there be pence, reckon each penny as the twelfth of £1, thus 1d. as 1s. 8d., and if there be farthings, each farthing as one-fourth the value of each penny, or 1q. as 5d. &c.

Ex. Find the value of 20 things at 2s. 8½d. each.

By the Rule,
The value required $= £1 \times 2 + (1s.\ 8d.) \times 8 + 5d. \times 2.$
$= £2 + 13s.\ 4d. + 10d. = £2.\ 14s.\ 2d.$

Reason for the Process.

20 things at $1s. = 20s. = £1$; ∴ 20 things at $2s.$
$= £1 \times 2 = £2,$
20 things at $1d. = 1s.\ 8d.$; ∴ 20 things at $8d.$
$= 1s.\ 8d. \times 8 = 13s.\ 4d.$
20 things at $\frac{1}{2}d. = \dfrac{1s.\ 8d.}{2}$, or 20 things at $\frac{1}{2}d. = 10d.$;

∴ value of 20 things at $2s.\ 8\frac{1}{2}d. = £2.\ 14s.\ 2d.$

156. *To find the value of 100 things, the value of one thing being given.*

RULE. Reckon each shilling in the given value as £5; reduce the pence and farthings in the given value to farthings, then reckon each farthing as equal to $2s.\ 1d.$

Ex. Find the value of 100 things at $2s.\ 5\frac{1}{4}d.$ each.
By the Rule, since $5\frac{1}{4}d. = 21q.$
The value req$^d. = £5 \times 2 + 2s. \times 21 + 1d. \times 21.$
$= £10 + £2.\ 2s. + 1s.\ 9d. = £12.\ 3s.\ 9d.$

Reason for the Process.

100 things at $1s. = £5$; ∴ 100 things at $2s. = £5 \times 2 = £10.$

Again since $1d. = 4q.$, taking $1q.$ as equal to $1d.$, we multiply by 4.

Also, since $2s. = 96q.$, taking $1q.$ as equal to $2s.$, we multiply by 96;

∴ taking $1q. = 2s. + 1d.$, we multiply by $96 + 4$, or 100.

157. *To find the interest of any sum of money for any number of months at 6 per cent.*

RULE. Divide the number of months by 2; the quotient is the interest in cents of $1 for the given time; multiply the quotient by the given principal and the product is the interest required.

Ex. 1. Find the interest on $78.56 for 2 yrs., 7 mo., at 6 per cent. per annum.

By the Rule,
2 yrs. 7 mo. $= 31$ months; $\frac{31}{2} = 15\frac{1}{2}.$
∴ int. req$^d. = 15\frac{1}{2} \times \$78.56 = \$12.1768.$

Reason for the Process.
The interest of $1 for 1 month = ½ cent.
∴ half the number of months will express the interest in cents of $1 for the given time.

Note 1. It will be quite easy to obtain from the above the interest at any other rate than 6 per cent.: by first obtaining the interest as directed above and then by Practice to add or subtract as the case may require.

Ex. 2. Find the interest of $80 for 15 months at 8 per cent. per annum.

At 6 per cent. int. = $6, as by the above Rule;
∴ at 8 per cent. int. = $6 + ⅓ of $6.
= $8.

Ex. 3. Find the interest on $110 for 10 months at 5 per cent. per annum.

At 6 per cent. int. = $5.50 by the Rule;
∴ at 5 per cent. int. = $5.50 − ⅙ $5.50.
= $5.50 − 91⅔ cents.
= $4.58⅓.

Note 2. If there are days in the question, the interest may be found for $1 by dividing the days by 6 and reckoning the quotient tenths of a cent, which being added to the result obtained in the Rule, will give the interest of $1 for months and days, and consequently for any amount.

Ex. 4. Find the interest on $90 for 6 months and 24 days at 6 per cent per annum.

Int. on $1 = 3·4 cents, by the Rule;
∴ int. on $90 = 3·4 cents × 90.
= $3.06.

Ex. LXXXI.

Find the interest at 6 per cent. per annum : (1) On $37 for 4 months. (2) On $42 for 6 months. (3) On $55 for 8 months. (4) On $75 for 10 months. (5) On $110 for 7 months. (6) On $38.50 for 9 months. (7) On $84.25 for 12 months. (8) On $120 for 15 months. (9) On $228 for 18 months. (10) On $678.50 for 8 months. **(11)** On $422.25 for 9 months. (12) On $328.50 for 9 months.

ANSWERS.

Ex. I. (p. 10.)

1. 3, 4, 2, 7, 9, 6, 8.   2. 10, 1, 12, 19, 5, 11, 16.   3. 14, 20, 27, 33, 49, 60, 55, 17, 36.   4. 88, 35, 63, 29, 76, 80, 94, 13, 52.   5. 9, 10, 11, 12, 13, 14, 15, 16, 17 ; 46, 47, 48, 49, 50, 88, 89, 90, 91, 92, 93, 94, 95, 96, 97, 98.

Ex. II. (p. 11.)

1. 106, 150, 200, 287, 310, 431, 555, 919, 867.
2. 196, 197, 198, 199, 200, 201, 202, 203, 204, 205, 206, 207, 208, 209, 210, 211, 212, 213 ; 612, 613, 614, 615, 616, 617, 618, 619 ; 948, 949, 950, 951, 952, 953, 954, 955, 956, 957, 958, 959, 960, 961, 962, 963, 964, 965, 966, 967, 968, 969.

Ex. III. (p. 12.)

1. 4585, 7321, 9798, 7006.
2. 5004, 5400, 5040, 8036, 8306, 8360, 9909.
3. 75635, 90909, 10004, 87050, 90001, 64064, 83000.

Ex. IV. (p. 13.)

1. 105, 8790, 37071, 30402, 77700, 24817.
2. 105409, 8008013, 7650090, 64000400, 89044001. 504623024, 900300800, 53000503.
3. 6006070007, 83401001010, 7004089200, 990000000.

Ex. V. (p. 14.)

1. Seven, thirteen, four, nine, eighteen, five, twenty, eleven, five, fifty, thirty-four, twenty-nine, three, seventeen, fifty-three.
2. Nineteen, eight, forty-one, eighty-eight, twenty-seven, seventy-two, ninety-four, forty-nine, sixteen, sixty-one, ninety-eight, eighty, fifty-six, twenty-eight.
3. One hundred and seven, one hundred and seventy, seventeen, four hundred and thirty, six hundred and ninety-one, eighty, eight hundred, eight, nine hundred and fifty-six, eight hundred and three, six hundred and eighty-four.
4. Four thousand five hundred and six, five thousand eight

ANSWERS.

hundred and seventy, five thousand and eighty-seven, six thousand nine hundred, six thousand and nine, two thousand five hundred and eighty, seven thousand and forty-five, seven thousand five hundred and ninety-one, six thousand two hundred and seventy-five.

5. Twenty-four thousand seven hundred and fourteen, twelve thousand five hundred, ten thousand and twenty-five, ten thousand two hundred and five, seventy thousand four hundred and fifty-seven, seventy-four thousand and seven, seventy-seven thousand.

6. Three hundred thousand eight hundred and sixty-three, thirty millions eighty thousand six hundred and thirty, ninety-six millions four hundred thousand two hundred and fifty, eight hundred millions four hundred thousand three hundred and seven, five hundred and seventy-two millions sixty thousand four hundred and ninety-five.

7. One hundred and twenty millions one hundred and ninety-two thousand seven hundred and three, eight hundred and ninety millions six hundred and forty-seven thousand five hundred and sixty, one billion and fifty millions sixty thousand four hundred and twenty-nine, one hundred billions and one.

Ex. VI. (p. 16.)

1. 19. 2. 27. 3. 26. 4. 11; 16; 18; 18; 23; 17; 15; 18; 25. 5. 25; 20; 34; 28; 36; 45; 46. 6. 29 boys. 7. 12 yrs. 8. 30 chickens.

Ex. VII. (p. 19.)

1. 37. 2. 69. 3. 99. 4. 99. 5. 95. 6. 71.
7. 115. 8. 110. 9. 200. 10. 214. 11. 213. 12. 186.
13. 214. 14. 241. 15. 503. 16. 1741. 17. 2133.
18. 1540. 19. 2201. 20. 1364. 21. 1920. 22. 1549.
23. 1551. 24. 2514. 25. 1665. 26. 2451. 27. 2148.
28. 2018. 29. 14658. 30. 27640. 31. 27832. 32. 35735.
33. 28260. 34. 29635. 35. 28207. 36. 100 marbles.
37. 257. 38. 9770. 39. 842068. 40. 3554 oranges.

Ex. VIII. (p. 20.)

1. 148. 2. 316. 3. 295. 4. 291. 5. 340. 6. 1851.
7. 2124. 8. 3216. 9. 3166. 10. 2974. 11. 336508.
12. 323915. 13. 400257. 14. 358064. 15. 152375.
16. 37155818. 17. 24601758. 18. 171357568. 19. 260342506.
20. 222275. 21. 186839. 22. 72268. 23. 194207.

ANSWERS.

Ex. IX. (p. 21.)

1. 2643560.   2. 5074223.   3. 226987.   4. 9948324.
5. 80169315.   6. 1642844613.   7. 5481487220.   8. 3582727022.
9. 5198944018.   10. 2553242166.   11. 4803131181.
12. 6137065059.   13. 434883345.   14. 100.   15. 982.
16. $3185.   17. 165802.

Ex. X. (p. 25.)

1. 4.   2. 12.   3. 28.   4. 50.   5. 26.   6. 546.   7. 156.
8. 6.   9. 2.   10. 58.   11. 36.   12. 9.   13. 16.   14. 35.
15. 184.   16. 167.   17. 188.   18. 198.   19. 601.
20. 594.   21. 205.   22. 87.   23. 89.   24. 109.   25. 179.
26. 98.   27. 109.   28. 398.   29. 13; 42; 38; 114; 198;
174.   30. $260; $40.   31. 19.   32. 31.   33. 8.   34. 7 cents.

Ex. XI. (p. 26.)

1. 1921.   2. 3711.   3. 999.   4. 2239.   5. 4484.   6. 1929.   7. 3205.   8. 4684.   9. 3401.   10. 7889.   11. 3025.
12. 806.   13. 25184.   14. 21023.   15. 8.   16. 18173.
17. 168079.   18. 8639494.   19. 19075299.   20. 555939946.
21. 2899; 997833.   22. 5986.   23. 15022.   24. 1891; 72.
25. 68; 140.   26. $217 in debt.   27. 19th step from bottom, 26th step from the top.   28. 5 officers, 58 children, 47 able-bodied, 23 infirm.   29. 682.   30. 45718.   31. 7096305.
32. 56289613.   33. 66291414.   34. $260.   35. $8337588.

Ex. XII. (p. 28.)

1. III; VII; XI; IX; XII; XVI; XVIII; XXV; XXVIII; XXXVII; XL; LIII; LIX; LXII; LXXVII; LXXXIV; CIII; CLVII; CXC; CC; DCLI; DCCLXXXIII; MCCIV; MDXXVII; MDCCCLXV.

2. three, 3; six, 6; eight, 8; thirteen, 13; fifteen, 15; seventeen, 17; twenty, 20; fifty-four, 54; eighty-one, 81; one hundred and eleven, 111; six hundred and five, 605; five thousand and two, 5002; one million one hundred thousand, 1100000; two thousand, 2000; seven hundred and forty-nine, 749; one thousand eight hundred and sixty-five, 1865.

Ex. XIII. (p. 30.)

1. 106.   2. 94.   3. 176.   4. 112.   5. 144.   6. 180.
7. 87.   8. 225.   9. 108.   10. 204.   11. 332.   12. 450.
13. 335.   14. 215.   15. 216.   16. 594.   17. 468.   18. 189.
19. 371.   20. 360.   21. 616.   22. 621.   23. 486.   24. 200.

ANSWERS. 175

25. 990.   26. 583.   27. 957.   28. 1001.   29. 720.   30. 588.
31. 1374.   32. 2400.   33. 2091.   34. 1104.   35. 3885.
36. 2982.   37. 3353.   38. 6335.   39. 6680.   40. 4383.
41. 5600.   42. 5918.   43. 10656.   44. 8448.   45. 429
bushels, 2574 shillings.   46. 756 pence, 1764 pence, 3024
pence.   47. 44 cents.   48. 44.   49. 885.

Ex. XIV. (p. 31.)

1. 18096.   2. 11698.   3. 29619.   4. 114228.   5. 24228.
6. 485340.   7. 416160.   8. 404825.   9. 3073630.   10. 388064.
11. 231483.   12. 346284.   13. 590592.   14. 833184.
15. 234927.   16. 1098444.   17. (1) 7740984, 19352460,
11611476, 27093444, 15481948, 34834428, 23222952, 30963936,
42575412, 46445904   (2) 9219516, 23048790, 13829274,
32268306, 18439032, 41487822, 27658548, 36878064, 50707338,
55317096.   (3) 171947728, 429869320, 257921592, 601817048,
343895456, 773764776, 515843184, 687790012, 945712504,
1031686368.   (4) 18181706, 45454265, 27272559, 63635971,
36363412, 81817677, 54545118, 72726824, 99999383, 109090236.
(5) 111760184, 279400460, 167640276, 391160644, 223526368,
502920828, 335280552, 447040736, 614681012, 670561104.
(6) 1975308642, 4938271605, 2962962963, 6913580247,
3950617284, 8888888889, 5925925926, 7901234568, 10864197531,
11851851852.   18. 98 miles.   19. 888 miles.

Ex. XV. (p. 34.)

1. 8334.   2. 18306.   3. 9108.   4. 32454.   5. 57706.
6. 32643.   7. 23790.   8. 22385.   9. 77341.   10. 42182.
11. 50516.   12. 79992.   13. 218075.   14. 281504.
15. 45468.   16. 303102.   17. 6964704.   18. 4328192.
19. 183150.   20. 331200.   21. 308163.   22. 250200.
23. 725912.   24. 1619723.   25. 52470000.   26. 492463028.
27. 7851033000.   28. 244366672.   29. 140645085.
30. 353446772.   31. 344115512.   32. 736924245.
33. 663503082.   34. 593928000000.   35. 8106030522.
36. 622439160.   37. 33146651 ; 1368500000 ; 791627400 ;
280812862751.5.   38. 148672.   39. (1) 61299 ; (2) 51480000.
40. See 15, 16, 17, 18.

Ex. XVI. (p. 34.)

1. 43042883.   2. 131296032.   3. 4916047312.   4. 43506216.
5. 31884470.   6. 88789980848.   7. 66260991808.
8. 40880656300.   9. 69312233476002.   10. 18381130075.
11. 100453365411.   12. 157593610868.   13. 8943214050.

L

ANSWERS.

14. 27416327796. 15. 109588282650. 16. 60435674536845.
17. 495562849756. 18. 67226401140. 19. 18834779670.

Ex. XVII. (p. 37.)

1. $14\frac{4}{5}$, $9\frac{7}{9}$, 11 ; $15\frac{5}{8}$, $10\frac{5}{9}$, $11\frac{5}{8}$; $16\frac{5}{8}$, $10\frac{8}{9}$, $12\frac{3}{8}$; $17\frac{1}{4}$, $11\frac{3}{8}$, $12\frac{7}{8}$; $16\frac{4}{5}$, $11\frac{1}{2}$, $12\frac{1}{3}$.

2. 21, $9\frac{5}{11}$, $10\frac{5}{13}$; 22, 10, 11 ; $23\frac{4}{5}$, $10\frac{8}{11}$, $11\frac{9}{13}$; $25\frac{3}{5}$, $11\frac{7}{11}$, $12\frac{8}{13}$; $23\frac{7}{8}$, $10\frac{7}{11}$, $11\frac{7}{13}$.

3. $21\frac{4}{5}$, $10\frac{10}{17}$, $11\frac{9}{11}$; $23\frac{3}{5}$, $11\frac{8}{13}$, $12\frac{9}{11}$; $25\frac{3}{5}$, $12\frac{9}{13}$, $13\frac{10}{11}$; 28, 14, $15\frac{3}{11}$; $24\frac{3}{5}$, $12\frac{3}{13}$, $13\frac{4}{11}$.

4. $28\frac{4}{5}$, $21\frac{4}{5}$, $14\frac{4}{17}$; $32\frac{3}{8}$, $24\frac{3}{8}$, $16\frac{3}{17}$; $34\frac{3}{5}$, $25\frac{4}{5}$, $17\frac{2}{17}$; $42\frac{3}{8}$, $32\frac{1}{8}$, $21\frac{5}{17}$; 40, 30, 20.

5. $115\frac{2}{3}$, $46\frac{2}{15}$, 42 ; $170\frac{7}{9}$, $68\frac{2}{15}$, 62 ; 210, 84, $76\frac{4}{11}$; $101\frac{1}{4}$, $40\frac{5}{15}$, $36\frac{8}{11}$; $138\frac{3}{4}$, $55\frac{3}{15}$, $50\frac{5}{11}$.

6. $54\frac{6}{11}$, 75, 50 ; $69\frac{4}{11}$, $95\frac{3}{8}$, $63\frac{7}{13}$; $76\frac{4}{11}$, $105\frac{3}{8}$, $70\frac{2}{13}$; $90\frac{9}{11}$, $124\frac{7}{8}$, $83\frac{3}{13}$; $65\frac{1}{11}$, $89\frac{3}{8}$, $59\frac{9}{13}$.

7. $134\frac{4}{5}$, $100\frac{10}{9}$, 110 ; 764, 573, $625\frac{1}{11}$; $784\frac{7}{8}$, $588\frac{7}{13}$, $642\frac{1}{11}$; $555\frac{8}{9}$, $416\frac{3}{13}$, $454\frac{6}{11}$.

8. 345, 230, 276 ; $1200\frac{4}{5}$, $800\frac{4}{17}$, $960\frac{4}{15}$; $1033\frac{3}{8}$, $688\frac{1}{12}$, $826\frac{7}{15}$; $818\frac{4}{5}$, $545\frac{8}{17}$, $654\frac{8}{15}$.

9. $7187\frac{2}{7}$, $8624\frac{8}{15}$, $12320\frac{4}{7}$; $6052\frac{1}{2}$, $7263\frac{8}{15}$, $10376\frac{4}{7}$; $7124\frac{2}{7}$, 8549, $12212\frac{4}{7}$; $2941\frac{4}{17}$, $3529\frac{8}{15}$, $5042\frac{4}{7}$.

10. $6909\frac{3}{11}$, $9500\frac{2}{3}$, $6333\frac{4}{13}$; $8182\frac{7}{11}$, $11251\frac{1}{8}$, $7590\frac{2}{13}$; $4820\frac{7}{11}$, $6628\frac{3}{8}$, $4418\frac{1}{2}$.

11. $683837\frac{2}{3}$, $547069\frac{8}{15}$, $497336\frac{2}{11}$; $1171 25 8 5\frac{2}{3}$, $9370068\frac{2}{15}$, $8518243\frac{2}{11}$; $2575524\frac{4}{5}$, $206019\frac{3}{15}$, $187290\frac{8}{11}$.

12 $1194292\frac{4}{7}$, $928894\frac{1}{2}$, $696670\frac{7}{13}$; 969949, $754404\frac{7}{9}$, $565803\frac{7}{13}$; $1412855\frac{4}{7}$, $1098887\frac{4}{9}$, $824165\frac{7}{13}$.

13. 66725 times. 19871. 14. (1) 9. (2) 1613. 15. 54 cents. 16. 7 plums. 17. 506. 18. 11946419. 19. Cook received $561, man-servant $1122, housekeeper $2244. 20. 1728. 21. 6. 22. 26 oranges. 23. 35 penknives.

Ex. XVIII. (p. 41.)

1. 12 ; 40 ; 53 ; 94. 2. 45 ; 29 ; 65 ; 97. 3. 57 ; 79 ; 88 ; 73. 4. 215 798 ; 885 ; 102. 5. 805 ; 682 ; 127 ; 357 ; 460 ; 7090. 6. 379 ; 407 ; 940 ; 738 ; 93845796 ; 580073. 7. 347 ; 569 ; 3094. 8. 1987 ; 7071 ; 650. 9. 9009 ; 5436 ; 388. 10. 21503 rem. 5 ; 3450 ; 124 rem. 477. 11. 57096 ; 76542 ; rem. 136 ; 4655 rem. 603. 12. 103944 ; 175971 rem. 66 ; 87039 ; 84003 ; 967427210 rem. 61. 13. 190182 ; 4623 ; 50301 ; 87366 rem. 6076. 14. 2007 rem. 1 ; 20300 : 65839 rem. 2 ; 31352. 15. 902468 ; 1754 rem 129 ; 14957000 ; 770071. 16. 37810 ; 3250450 ; 73086413. 17. 1799. 18. 180 pairs.

ANSWERS. 177

19. 141. **20.** 360 rem. 52. **21.** $3½⅞. **22.** $3. **23.** 1000.
24. 420. **25.** 403. **26.** 372547. **27.** 17129. **28.** $10.

Ex. XIX. (p. 43.)

1. $3\frac{1}{10}$, $4\frac{3}{10}$, $5\frac{9}{10}$, 8, $13\frac{5}{10}$, 26, $150\frac{1}{10}$; $1\frac{11}{20}$, $2\frac{3}{20}$, $2\frac{19}{20}$, 4, $6\frac{15}{20}$, 13, $75\frac{4}{20}$; $1\frac{4}{30}$, $1\frac{18}{30}$, $1\frac{28}{30}$, $2\frac{23}{30}$, $4\frac{13}{30}$, $8\frac{28}{30}$, $50\frac{5}{30}$.

2. $5\frac{37}{40}$, 21, $16\frac{33}{40}$, $7\frac{11}{40}$, $150\frac{19}{40}$, $195\frac{29}{40}$, $2030\frac{29}{40}$, 8195; $3\frac{57}{70}$, 14, $11\frac{13}{70}$, $4\frac{61}{70}$, $100\frac{19}{70}$, $130\frac{29}{70}$, $1353\frac{49}{70}$, $5463\frac{29}{70}$; $3\frac{27}{70}$, 12, $9\frac{43}{70}$, $4\frac{11}{70}$, $85\frac{69}{100}$, $111\frac{59}{100}$, $1160\frac{39}{100}$, $468\frac{299}{700}$; $2\frac{37}{100}$, $8\frac{49}{100}$, $6\frac{73}{100}$, $2\frac{91}{100}$, $60\frac{19}{100}$, $78\frac{29}{100}$, $812\frac{29}{100}$, 3278; $1\frac{37}{200}$, $4\frac{49}{200}$, $3\frac{73}{200}$, $1\frac{91}{200}$, $30\frac{19}{200}$, $39\frac{29}{200}$, $406\frac{29}{200}$, 1639.

3. $329\frac{38}{270}$, $281\frac{49}{240}$, $3708\frac{141}{270}$; $79\frac{188}{1000}$, $6\frac{870}{1000}$, $890\frac{81}{1000}$; $521\frac{948}{1300}$, $4\frac{878}{1000}$, $593\frac{661}{1500}$; $30\frac{1043}{2800}$, $21\frac{679}{2300}$, $342\frac{941}{2600}$; $8306\frac{781031490}{100000000}$.

4. $88976\frac{858}{6009}$. 5. $86782\frac{4919}{10000}$. 6. $66970\frac{719}{809}$. 7. $7096\frac{8884}{9586}$.
8. $9992461\frac{298}{585}$. 9. $2144\frac{8591}{5325}$. 10. $1302\frac{837}{8087}$. 11. 1100, 916 and 800 men over. 12. $956\frac{9299}{100000}$.

Ex. XX. (p. 51.)

1. 681440 far. 2. 1085070 inches. 3. 4167680 drs.
4. 3842027640 sq in. 5. 8092505 ls. 6. 31518396 sq. in.
7. 24480 mins. 8. 16820 grs. 9. 15620 yds.
10. 1074088 c. in. 11. 440 gills. 12. 7040 qts.
13. 2030400 mins. 14. 158304 grs. 15. 276400 grs.
16. 96425 half-pence. 17. 1062864 sq. yds. 18. 3499 nls.
19. 2281906 far. 20. 21667 lbs. 21. 92160 secs.
22. 530784 in. 23. 300362 far. 24. 604800 grs. 25. 520 nls.
26. 888 nls. 27. 544345 far. 28. 82800 scs. 29. 378 galls.
30. 192938 far.

Ex. XXI. (p. 53.)

1. £128 8s. 6½d. 2. 2 lbs. 3 oz. 8 dwt. 20 grs.
3. 2273 galls. 3 qts. 1 pt. 4. 403 lea. 2 mls. 7 fur. 16 po.
5. 3 tons 18 cwt. 1 qr. 14 lbs. 14 oz. 6. 586 ac. 1 ro. 27 po.
7. 29 lbs. 1 oz. 12 dwt. 4 grs.
8. 11517 mls. 1 fur. 27 po. 2 yds. 11 in. 9 ls.
9. 14997 tons 8 cwt. 1 qr. 14 lbs. 10 oz. 12 drs.
10. 1 ml. 7 fur. 14 po. 2 ft. 9 in. 11. 17 lbs. 3 dwt. 14 grs.
12. 3 tons 19 cwt. 1 lb. 6 oz. 13. 122 lbs. 3 drs. 17 grs.
14. 2 wks. 5 dys. 23 hrs. 58′ 13″. 15. 35 ac. 2 ro. 20 po.
16. 297 c. yds. 17. 198 ac. 1 ro. 15 po. 16¼ yds.
18. 31 yds. 1 qr. 19. 36° 24′ 35″. 20. 365 dys. 6 hrs.
21. 508 hhds. 19 gals. 2 qts. 22. 596 hhds. 14 gals. 1 qt.
23. 15211 bu. 55 lbs. 24. 29411 bu. 26 lbs. 25. 121 bu. 3 lbs.
26. $307.47. 27. £1014 4s. 3¾d.

ANSWERS.

Ex. XXII. (p. 54.)

1. $94.64.   2. £20. 12s. 3l.   3. 10 qrs., 24 lbs., 1 oz.
4. 107 lbs., 1 oz., 10 dwt., 17 grs.
5. 55 lbs., 1 oz., 5 drs., 2 sc., 1 gr.
6. $17255.22.   7. 288 tons, 2 cwt., 2 qrs., 23 lbs.
8. 578 yds., 2 qrs.   9. 79 mls., 3 fur., 9 per., 3 yds.
10. £145. 17s. 1½d.   11. 116 dys., 8 hrs., 35′, 12″.   12. $8470.12.
13. 42 ac., 2 ro., 25 po., 5 ft., 40 in.
14. 99 tons, 8 cwt., 3 qrs., 12 lbs., 11 oz., 15 drs.
15. $11040.

Ex. XXIII. (p. 56.)

1. £15. 8s. 6d.   2. 9 lbs., 11 oz., 3 drs., 16 grs.
3. 2 lbs., 10 oz., 7 dwt.   4. 2 mls., 6 fur., 35 po., 1 yd.
5. 13 yds., 1 qr., 2 nls., 2 in.   6. 28 c. yds., 23 c. ft., 1443 in.
7. 1 ac., 2 ro., 38 po., 1 yd., 2 ft., 142 in.
8. 5 dys., 9 hrs., 49 min., 57 sec.   9. £53. 17s. 10¾d.
10. 2 qrs., 15 lbs., 11 oz., 14 drs.   11. $1068.89.
12. 95 cords, 110 c. ft.   13. $27.69.
14. 107 ac., 2 ro., 34 po., 29 yds., 7 ft., 118 in.
15. 79 c. yds., 21 c. ft., 1377 c. in.
16. 27 mls., 29 per., 1 ft., 10 in.   17. 6°, 39′, 39″.
18. 5 tons, 16 cwt., 2 qrs., 23 lbs., 11 oz., 1 dr.
19. 10 yds., 2 qrs., 2 nls., 2 in.   20. 70 bu., 2 pks., 1 gal., 2 qts.
21. 673 bu., 1 gal., 2 qts.

Ex. XXIV. (p. 58.)

1. £24. 19s.   2. 52 lbs., 5 oz., 4 drs.
3. 74 lbs., 1 oz., 1 dwt., 16 grs.   4. 139 yds., 2 qrs., 3 nls.
5. 167 mls., 6 fur., 1 per., ½ yd.   6. $1660.33.
7. 129 cwt., 1 qr., 11 lbs., 7 oz., 8 drs.
8. 58 mls., 5 fur., 18 po., 1 yd., 9 in.   9. $6099.30.
10. 86 wks., 8 hrs., 56 min.   11. 95 ac., 36 per., 3 ft.
12. £146. 3s. 6¼d.   13. 899 lbs., 8 oz., 4 drs.
14. 23 bu., 1 pk., 3 qts.   15. 21 dys., 15 hrs., 50 min.
16. 50103 gals., 2 qts., 1 pt.
17. 4382 dys., 21 hrs., 47 min., 24 sec.   18. £812. 15s. 0¼d.
19. 134 ac., 3 ro., 31 po.   20. $3572.16.
21. 25043 bu., 2 pks., 1 gal.   22. £840. 11s. 6d.
23. 219 lbs., 8 oz., 10 dwt., 12 grs.   24. $7342.
25. £159. 15s. 6¾d.   26. $22503.   27. 381 mls., 12 po., 2 yds.

Ex. XXV. (p. 59.)

1. 1583 ac., 2 ro., 12 po.   2. 1500 mls., 6 po.   3. £2817. 12s.
4. 1621 lbs., 4 oz., 15 dwt., 13 grs.   5. £351. 13s. 9d.

ANSWERS. 179

6. 1484 yds., 2 qrs., 2 nls. 7. 188 cwt., 22 lbs., 11 oz., 10 drs.
8. £5912. 4s. 9¾d. 9. $7321.30.
10. 1493 c. yds., 11 c. ft., 1332 in.
11. 182 lbs., 10 oz., 1 dwt., 13 grs. 12. £3743. 7s. 10d.
13. 688 dys., 6 hrs., 40 min. 14. 6297 lbs., 11 oz., 4 drs.
15. 33272 lbs., 1 oz., 18 dwt., 6 grs. 16. £3676. 13s. 10¼d.
17. 1319 ac., 0 ro., 0 po., 13 yds., 4 ft., 48 in.
18. 1034 mls., 2 fur., 4 po., 3 in. 19. £2100. 18s. 9d.
20. $118575. 21. 8500 bushels.

Ex. XXVI. (p. 61.)

1. 352 cwt., 2 qrs., 21 lbs., 13 oz.
2. 33772 lbs., 10 oz., 18 dwt., 15 grs. 3. £2194. 10s. 7d.
4. 1870 cwt., 3 qrs., 23 lbs., 4 oz., 5 drs. 5. £2771. 2s. 1½d.
6. 10826 lbs., 8 oz., 5 drs., 2 sc., 4 grs. 7. $470.25.
8. $66.04. 9. $97.50. 10. 12 cwt., 1 qr., 7 lbs., 8 oz.
11. $40000.

Ex. XXVII. (p. 63.)

1. £55. 15s. 1¾d. 2. 29 lbs., 8 oz., 3 dwt., 6 grs.
3. 24 mis. 1 fur., 19 po., 3 yds., 10 in. 4. 23 yds., 2 nls.
5. 144 lbs., 3 oz., 4 drs., 8 ½ grs. 6. £188. 19s. 9¾d.
7. 2 tons, 10 cwt., 12 lbs., 10 oz., 10⅜ drs.
8. 17 ac., 1 ro., 30 po., 10 yds., 6 ft., 55$\frac{11}{17}$ in.
9. 3 qrs., 11 lbs., 13 oz., 5$\frac{10}{17}$ drs.
10. 6 mls, 7 fur., 14 po., 3$\frac{617}{717}$ in.
11. 6 bu., 1 pk., 1 gal., $\frac{11}{17}$ pt. 12. £52. 16s. 2¾d.
13. 4 lbs., 4 oz., 1 dr., 1 scr., 11$\frac{13}{17}$ grs.
14. 7 fur., 23 po., 5 yds., 1 in., 1$\frac{55}{211}$ ls.
15. 1 ac., 1 ro., 9 po., 22 yds., 5 ft., 14$\frac{102}{311}$ in.
16. 1 ton, 1 cwt., 3 qrs., 2 lbs., 12 oz., 10$\frac{331}{354}$ drs.
17. 5 c. yds., 11 c. ft., 961$\frac{33}{64}$ c. in.
18. 4 lbs., 4 oz., 3 drs., 1 sc., 11$\frac{17}{21}$ grs.
19. 4 lbs. 10 oz., 1 dwt., 9$\frac{217}{231}$ grs. 20. £2. 10s. 6$\frac{117}{154}$d.
21. 13s. 7¼d. 22. 115 dys., 5 hrs., 54 min., 22$\frac{34}{59}$ sec.
23. $8 49$\frac{101}{201}$. 24. 13.39\frac{299}{399}$. 25. 1.64\frac{110}{210}$. 1.91\frac{217}{315}$.
26. $4.65. 27. $1.15. 28. 2.53\frac{11}{44}$.

Ex. XXVIII. (p. 64.)

1. 9 times. 2. 3 times. 3. 436 times. 4. 3 times.
5. 25$\frac{125}{133}$ times. 6. 8$\frac{233}{253}$ times. 7. 9 times. 8. 24 times.
9. 75 times. 10. 65 times. 11. 100 times.

Ex. XXIX. (p. 65.)

1. $101.25. 2. $231.85. 3. $831.53½. 4. $615 68¼

ANSWERS.

5. $871.40. 6. $1279.12½. 7. $2451.98¾. 8. $253.96¼.
9. $3650.50. 10. $2845.09⅛. 11. $4468.12½. 12. $189.55.
13. $8069.33⅛. 14. $301.94⅛. 15. $151.72½. 16. $350.75.

Ex. XXX. (p. 66.)

1. £54. 6s. 3d.   2. £81. 17s. 9d.   3. £4. 6s. 9d.
4. £21. 2s. 6d.   5. £18. 19s. 9d.   6. £31. 6s. 10½d.
7. £216. 19s. 4½d.   8. £290. 12s.   9. £348. 8s. 1½d.
10. £484. 6s.   11. £555. 1s. 5⅝d.   12. £946. 7s. 4⅛d.

Ex. XXXI. (p. 66.)

PAPER I.

1. 117984.   2. 107766.   3. 3653012.   4. 1898307.
5. 2 mls., 6 fur., 1 per., 5 yds., 1 ft., 10 in.
6. 1st, $5.60, $17.17 ; 2nd, $11.57.

PAPER II.

1. £362. 19s. 9d.   2. $63.47.
3. 183 ac., 1 ro., 24 per., 26 yds., 7¹⁵⁵⁄₃₅₇ ft.
4. $9.48⁴⁄₃₇.   5. 5 dresses, £2. 15s. 7¼d. each.   6. $3227.42.

PAPER III.

1. £46. 14s. 6d.   2. $7000, $21000, $35000.   3. £13. 12s. 9d.
4. 1 ro., 18 po., 5 yds., 2 ft., and 16¼ ft. over.
5. 17 cwt., 1 qr., 8 lbs., 10 oz., 5 drs., and 89 drs. over.
6. 6 hours, 54 min.

PAPER IV.

1. $148.15.   2. 11, 18.   3. 9, 18, 27.   4. owner of net, 8 dozen ; owner of boat, 16 dozen ; each man, 32 dozen.
5. 2301696 pores.   6. 42000, 42889.

PAPER V.

1. 6255647664981.   2. 861447920.   3. 11904.   4. 12752.
5. 465335.   6. 95587.

PAPER VI.

1. 657872.   2. $16496471.   3. $10444830.63.   4. 6228¼ lbs.
5. 136 ac., 3 ro., 14 po., 24⁵⁷⁴⁄₁₂₇₀ yds.   6. 634338.

Ex. XXXII. (p. 71.)

1. 2.   2. 3.   3. 2.   4. 4.   5. 4.   6. 3.   7. 2.   8. 6.   9. 4.
10. 2.   11. 58.   12. 63.   13. 2.   14. 30.   15. 10.   16. 8.
17. none.   18. 8.   19. 26.   20. 352.   21. 131.   22. none.
23. 7056.   24. 11.   25. 17.   26. 31.

ANSWERS.

Ex. XXXIII. (p. 72.)

1. 20. 2. 72. 3. 144. 4. 1260. 5. 240. 6. 168. 7. 525.
8. 1056. 9. 1050. 10. 2520. 11. 11088. 12. 450. 13. 1080.
14. 840. 15. 840. 16. 16380. 17. 1386. 18. 21000.
19. 43890. 20. 95040.

Ex. XXXIV. (p. 74.)

(1) $\dfrac{4}{7}, \dfrac{6}{7}, \dfrac{10}{7}, \dfrac{14}{7}, \dfrac{18}{7}, \dfrac{24}{7}; \dfrac{34}{19}, \dfrac{51}{19}, \dfrac{85}{19},$

$\dfrac{119}{19}, \dfrac{153}{19}, \dfrac{204}{19}.$

(2) $\dfrac{378}{84}, \dfrac{504}{84}, \dfrac{693}{84}, \dfrac{6678}{84}, \dfrac{9891}{84}; \dfrac{570}{107}, \dfrac{760}{107},$

$\dfrac{1045}{107}, \dfrac{10070}{107}, \dfrac{14915}{107}.$

Ex. XXXV. (p. 75.)

(1) $\dfrac{3}{8}, \dfrac{3}{12}, \dfrac{3}{20}, \dfrac{3}{24}, \dfrac{3}{36}, \dfrac{3}{48}; \dfrac{7}{18}, \dfrac{7}{27}, \dfrac{7}{45},$

$\dfrac{7}{54}, \dfrac{7}{81}, \dfrac{7}{108}.$

(2) $\dfrac{16}{87}, \dfrac{16}{145}, \dfrac{16}{319}, \dfrac{16}{1624}, \dfrac{16}{2900}; \dfrac{77}{267}, \dfrac{77}{445},$

$\dfrac{77}{979}, \dfrac{77}{4984}, \dfrac{77}{8900}.$

Ex. XXXVI. (p. 75.)

(1) $\dfrac{6}{2}, \dfrac{27}{9}, \dfrac{39}{13}; \dfrac{10}{2}, \dfrac{45}{9}, \dfrac{65}{13}; \dfrac{16}{2}, \dfrac{72}{9}, \dfrac{104}{13};$

$\dfrac{30}{2}, \dfrac{135}{9}, \dfrac{195}{13}.$

(2) $\dfrac{72}{8}, \dfrac{90}{10}, \dfrac{514}{57}; \dfrac{96}{8}, \dfrac{120}{10}, \dfrac{684}{57}; \dfrac{136}{8}, \dfrac{170}{10},$

$\dfrac{969}{57}; \dfrac{296}{8}, \dfrac{370}{10}, \dfrac{2109}{57}.$

Ex. XXXVII. (p. 76.)

1. 3. 2. $2\tfrac{1}{2}$. 3. $4\tfrac{1}{3}$. 4. 4. 5. $3\tfrac{1}{5}$. 6. $6\tfrac{5}{7}$. 7. $5\tfrac{8}{9}$. 8. $6\tfrac{10}{11}$.
9. 7. 10. 8. 11. $8\tfrac{11}{14}$. 12. $18\tfrac{19}{24}$. 13. $9\tfrac{87}{107}$. 14. $102\tfrac{20}{243}$.
15. $12\tfrac{88}{99}$.

Ex. XXXVIII. (p. 76.)

1. $\dfrac{4}{3}$ 2. $\dfrac{25}{12}$ 3. $\dfrac{16}{15}$ 4. $\dfrac{55}{5}$ 5. $\dfrac{58}{7}$ 6. $\dfrac{3874}{19}$

7. $\dfrac{141}{63}$ 8. $\dfrac{239}{8}$ 9. $\dfrac{88716}{126}$ 10. $\dfrac{360931}{401}$ 11. $\dfrac{3407}{680}$

12. $\dfrac{3376}{63}$ 13. $\dfrac{26253}{1250}$ 14. $\dfrac{69057}{465}$ 15. $\dfrac{29160}{2160}$

16. $\dfrac{60389}{2400}$ 17. $\dfrac{608543}{3084}$

Ex. XXXIX. (p. 77.)

1. $\dfrac{3}{5}$ 2. $\dfrac{2}{3}$ 3. $\dfrac{9}{19}$ 4. $\dfrac{12}{55}$ 5. $\dfrac{35}{16}$ 6. $\dfrac{5}{6}$ 7. $\dfrac{5945}{6}$

8. $\dfrac{3363}{35}$ 9. $\dfrac{35}{2}$ 10. $\dfrac{15}{2}$ 11. $\dfrac{1}{36}$ 12. $\dfrac{1}{11}$ 13. $\dfrac{375}{44}$

14. $\dfrac{175}{8}$ 15. $\dfrac{14}{15}$ 16. $\dfrac{6399}{22}$

Ex. XL. (p. 78.)

1. $\dfrac{1}{2}$ 2. $\dfrac{2}{3}$ 3. $\dfrac{2}{3}$ 4. $\dfrac{5}{8}$ 5. $\dfrac{4}{9}$ 6. $\dfrac{16}{21}$ 7. $\dfrac{7}{11}$

8. $\dfrac{3}{17}$ 9. $\dfrac{7}{13}$ 10. $\dfrac{11}{13}$ 11. $\dfrac{7}{8}$ 12. $\dfrac{13}{20}$ 13. $\dfrac{31}{84}$

14. $\dfrac{3}{4}$ 15. $\dfrac{35}{114}$ 16. $\dfrac{8}{9}$ 17. $\dfrac{191}{279}$ 18. $\dfrac{827}{7337}$

19. $\dfrac{235}{397}$ 20. $\dfrac{103}{136}$ 21. $\dfrac{2}{7}$ 22. $\dfrac{945}{1529}$ 23. $\dfrac{20}{21}$

24. $\dfrac{23}{33}$

Ex. XLI. (p. 79.)

1. $\dfrac{9}{12}, \dfrac{10}{12}$ 2. $\dfrac{9}{12}, \dfrac{8}{12}$ 3. $\dfrac{6}{8}, \dfrac{7}{8}$ 4. $\dfrac{27}{63}, \dfrac{35}{63}$ 5. $\dfrac{33}{48}, \dfrac{42}{48}$

6. $\dfrac{110}{120}, \dfrac{81}{120}$ 7. $\dfrac{140}{200}, \dfrac{183}{200}$ 8. $\dfrac{2712}{6720}, \dfrac{3689}{6720}$ 9. $\dfrac{48}{60}$,

$\dfrac{55}{60}, \dfrac{9}{60}$ 10. $\dfrac{189}{1008}, \dfrac{384}{1008}, \dfrac{560}{1008}$ 11. $\dfrac{98}{210}, \dfrac{110}{210}$,

$\dfrac{161}{210}$ 12. $\dfrac{6545}{8415}, \dfrac{6120}{8415}, \dfrac{7293}{8415}, \dfrac{4455}{8415}$ 13. $\dfrac{1170}{1260}$,

ANSWERS.

$\dfrac{1225}{1260}, \dfrac{1176}{1260}, \dfrac{800}{1260}.$ 1. $\dfrac{105}{180}, \dfrac{102}{180}, \dfrac{130}{180}, \dfrac{135}{180}, \dfrac{84}{180}.$

15. $\dfrac{3744}{7200}, \dfrac{6075}{7200}, \dfrac{4200}{7200}, \dfrac{6000}{7200}.$ 16. $\dfrac{399330}{621180}, \dfrac{448630}{621180},$

$\dfrac{149940}{621180}, \dfrac{319464}{621180}, \dfrac{340170}{621180}, \dfrac{484155}{621180}.$ 17. $\dfrac{80}{120}, \dfrac{96}{120},$

$\dfrac{45}{120}, \dfrac{112}{120}.$ 18. $\dfrac{90}{120}, \dfrac{100}{120}, \dfrac{105}{120}, \dfrac{108}{120}.$ 19. $\dfrac{48}{72}, \dfrac{60}{72},$

$\dfrac{63}{72}, \dfrac{40}{72}, \dfrac{45}{72}.$ 20. $\dfrac{220}{330}, \dfrac{264}{330}, \dfrac{165}{330}, \dfrac{90}{330}.$

Ex. XLII. (p. 80.)

1. $\dfrac{10}{15}, \dfrac{12}{15}.$ 2. $\dfrac{28}{36}, \dfrac{27}{36}.$ 3. $\dfrac{221}{312}, \dfrac{228}{312}.$ 4. $\dfrac{392}{504}, \dfrac{315}{504},$

$\dfrac{396}{504}.$ 5. $\dfrac{315}{882}, \dfrac{396}{882}, \dfrac{672}{882}.$ 6. $\dfrac{3339}{5040}, \dfrac{3528}{5040}, \dfrac{3420}{5040}.$

7. $\dfrac{40}{168}, \dfrac{36}{168}, \dfrac{1239}{168}.$ 8. $\dfrac{220}{300}, \dfrac{255}{300}, \dfrac{252}{300}, \dfrac{290}{300}.$ 9. $\dfrac{16704}{6336},$

$\dfrac{1728}{6336}, \dfrac{5445}{6336}, \dfrac{103040}{6336}.$ 10. $\dfrac{28}{63}, \dfrac{99}{63}, \dfrac{420}{63}.$

11. $\tfrac{4}{7}$ yd. is greater by $\tfrac{4}{35}$ yd. 12. $\tfrac{3}{8}$ yd. is greater by $\tfrac{1}{8}$ yd.
13. $1\tfrac{1}{8}$ of $\tfrac{4}{5}$ of $1\tfrac{2}{3}$ of $\tfrac{11}{4}$ of a loaf is greater by $\tfrac{9}{24}$ of a loaf.

Ex. XLIII. (p. 82.)

1. $\dfrac{13}{21}.$ 2. $1\tfrac{4}{17}.$ 3. $3\tfrac{1}{3}.$ 4. $1\tfrac{7}{12}.$ 5. $\dfrac{53}{60}.$ 6. $1\tfrac{1}{8}.$ 7. $1\tfrac{4}{11}.$

8. $\dfrac{41}{56}.$ 9. $\dfrac{37}{90}.$ 10. $2\tfrac{1}{2}.$ 11. $15\tfrac{1}{2}.$ 12. $9\tfrac{7}{12}.$ 13. $2\tfrac{2}{15}.$

14. $5\tfrac{59}{72}.$ 15. $9\tfrac{89}{128}.$ 16. $22\tfrac{26}{27}.$ 17. $2\tfrac{8}{13}.$ 18. $2\tfrac{49}{60}.$ 19. $15\tfrac{17}{60}.$
20. $21\tfrac{737}{440}.$ 21. $46\tfrac{293}{360}.$ 22. $29\tfrac{44}{77}.$ 23. $116\tfrac{11}{26}.$ 24. $£111\tfrac{113}{144}.$
25. $13\tfrac{49}{135}$ lbs.

Ex. XLIV. (p. 83.)

1. $\dfrac{1}{20}.$ 2. $\dfrac{3}{8}.$ 3. $\dfrac{1}{2}.$ 4. $\dfrac{5}{72}.$ 5. $1\tfrac{1}{2}.$ 6. $4\tfrac{1}{10}.$ 7. $1\tfrac{3}{8}.$

8. $3\tfrac{1}{4}.$ 9. $\dfrac{41}{100}.$ 10. $1\tfrac{61}{140}.$ 11. $7\tfrac{11}{21}.$ 12. $12\tfrac{113}{131}.$ 13. $\tfrac{1}{2}$ of

cake. 14. $(1)\dfrac{8}{9}$ $(2) 5\tfrac{7}{13}.$ 15. $\dfrac{5}{72} d.$

Ex. XLV. (p. 84.)

1. $12\frac{118}{210}$. 2. $\frac{1}{6}$. 3. $20\frac{2}{17}$. 4. $36\frac{2}{18}$. 5. 1. 6. $1\frac{13}{17}$. 7. $3\frac{13}{20}$.

8. $B, C, D,$ and A had respectively $\frac{1}{4}, \frac{1}{6}, \frac{1}{9},$ and $\frac{2}{9}$ of cheese.

Ex. XLVI. (p. 86.)

1. $\frac{1}{12}$. 2. $\frac{35}{72}$. 3. $\frac{5}{26}$. 4. $\frac{1}{12}$. 5. 25. 6. $2\frac{5}{8}$. 7. $\frac{13}{40}$.

8. 10. 9. $4\frac{2}{7}$. 10. $329\frac{1}{16}$. 11. $4\frac{2}{7}$. 12. $6\frac{11}{16}$. 13. $\frac{17}{32}$.

14. $5\frac{7}{38}$. 15. $7\frac{1}{2}$.

Ex. XLVII. (p. 86.)

1. $1\frac{7}{10}$. 2. $\frac{2}{3}$. 3. $\frac{10}{11}$. 4. $\frac{116}{165}$. 5. $9\frac{1}{4}$. 6. $1\frac{14}{15}$. 7. $\frac{2}{3}$.

8. $1\frac{74}{85}$. 9. $3\frac{111}{112}$. 10. $\frac{27}{88}$. 11. $\frac{1}{32}$. 12. $\frac{3}{8}$. 13. $\frac{13}{15}$.

Ex. XLVIII. (p. 87.)

1. $1\frac{81}{85}$. 2. $2\frac{1}{4}$. 3. $\frac{3}{8}$. 4. $1\frac{2}{5}$. 5. $1\frac{17}{11}$. 6. $\frac{28}{725}$. 7. $2\frac{1}{4}$,

8. $\frac{76}{153}$. 9. $12\frac{11}{13}$. 10. $1\frac{227}{1414}$. 11. $\frac{8}{9}$. 12. $\frac{810}{102949}$.

13. $\frac{25}{144}$. 14. $3\frac{113}{878}$.

Ex. XLIX. (p. 88.)

1. 40 cents. 2. 3 fur. 3. 1 qr., 17 lbs., 13 oz., $11\frac{3}{8}$ drs.
4. 19 cwt., 1 qr., 10 lbs. 5. 4 fur., 35 per. 6. 2 ac., 1 ro.,
25 per., 20 yds., 4 ft., $136\frac{3}{4}$ in. 7. 4 lbs., 2 oz., 10 dwt. 20 grs.
8. 59 yds., 2 qrs., $1\frac{3}{4}$ nls. 9. £7. 4s. 3d. 10. 109 lbs., 8 oz.,
5 drs. $8\frac{1}{2}$ grs. 11. 5 hrs., 36 min. 12. 7 lbs., 9 oz., $9\frac{3}{4}$ drs.
13. $24. 14. 7 hrs. 12 min. 15. 13 cords, 64 c. ft.

Ex. L. (p. 88.)

1. $\frac{1}{6}$. 2. $\frac{31}{160}$. 3. $\frac{15128}{15}$. 4. $\frac{263}{430}$. 5. $\frac{408}{577}$. 6. $\frac{175}{44}$.

ANSWERS.

7. $\frac{1}{45}$. 8. $\frac{19}{70}$. 9. $\frac{6}{11}$. 10. $\frac{1}{27}$. 11. $\frac{1}{28}$. 12. $\frac{144}{175}$.

13. $\frac{3}{224}$. 14. $\frac{3}{14960}$. 15. $\frac{325}{7850601}$.

Ex. LI. (p. 89.)

1. $\frac{1}{2}$. 2. $\frac{1080}{241}$. 3. $\frac{5445}{5762}$. 4. $\frac{357}{160}$. 5. $\frac{7}{4}$. 6. $\frac{56}{3}$.

7. $\frac{18}{35}$. 8. $\frac{21}{2}$. 9. $\frac{27}{14}$. 10. $\frac{156}{5}$. 11. $\frac{21625}{432}$.

12. $\frac{8}{7}$. 13. $\frac{351}{224}$. 14. $\frac{115}{96}$.

Ex. LII. (p. 92.)

1. A will have $\frac{1}{8}$ of the farm, B $\frac{1}{8}$ of farm, and C $\frac{3}{4}$ of farm. 2. (1) 24 boys; (2) $7\frac{1}{2}$. 3. $\frac{3}{4}$ and $\frac{1}{10}$. 4. (1) $1\frac{1}{5}$; (2) $2\frac{2}{3}$. 5. A has twice as much as D. 6. $25.20. 7. $110. 8. $900. 9. $36. 10. £70. 11. $385\frac{5}{15}$ rounds. 12. $33.60. 13. $13\frac{1}{9}$ days. 14. (1) A has $56.70, B has $37.80; (2) A has $63, B has $31.50. 15. $\frac{3}{4}$, $255.10\frac{1}{2}$. 16. $\frac{1}{2}$. 17. 250 boys. 18. $5\frac{1}{5}$ days. 19. Elder son received $3250, younger son $1560, and widow $1440. 20. A has 24 ac., 3 ro.; B has 13 ac., 2 ro.; and C has 47 ac., 1 ro. 21. 48 boys. 22. 42 min. 23. (1) $52\frac{1}{2}$ days; (2) $\frac{1}{2}$. 24. $15\frac{1}{2}$ days. 25. 1st cask contains 140 gals.; 2d, 60 gals.; 3d, 45 gals.; 4th, 80 gals. 26. 20 days.

Ex. LIII. (p. 95.)

1. $\frac{3}{10}$; $\frac{13}{100}$; $\frac{19}{100}$; $\frac{301}{1000}$; $\frac{270}{1000}$; 2. $\frac{504}{1000}$;

$\frac{73201}{100000}$; $\frac{791003}{1000000}$; $\frac{3}{100}$; $\frac{45}{10000}$. 3. $\frac{300}{1000}$; $\frac{18741}{1000}$; $\frac{21}{10}$;

$\frac{1}{1000000}$; $\frac{50007}{10000}$. 4. $\frac{34702007}{100000}$; $\frac{500005}{1000}$; $\frac{560746805}{100000000}$;

$\frac{500}{10000000}$. 5. $\frac{290050}{10000}$; $\frac{20607}{1000}$; $\frac{500038}{100000}$.

Ex. LIV. (p. 96.)

1. ·4; 2·3; 23·5; ·04; ·147; ·047. 2. 500·1; 9·51; ·981; 5·02; ·00502. 3. 35·6; 17·00701; ·0050005; ·0000002;

20·76854; ·0000053052.   4. ·7; ·030.   5. 300·003; ·0001.
6. 4·000504; ·0000070.

7. Six tenths; seventeen hundredths; seven hundredths.

8. Seven **thousandths**; seven hundred thousandths, or seven **tenths**; **six and three** thousand and four ten thousandths.

9. Thirty-five **and two** hundred and five hundred **thousandths**; four **hundred and thirty-four** thousand one **hundred** thousandths, **or four hundred and** thirty-four hundredths.

10. 3, 30, 3000, 3000000; 1·3, 13, 1300, 1300000; 540·003, 5400·03, ·540003, **540003000**; 74201, 742010, 74201000, 74201000000.   11. ·5362, ·05362, ·000005362; ·03, ·003, ·0000003; 7·00107, ·700107, ·000700107; 500, 50, ·005.
12. ·00000203.

Ex. LV. (p. 97.)

1. 560·34603.   2. 214·08691.   3. 10061·33654.
4. 345·608037.   5. 40·23111.   6. 585·07805.   7. 7332·0773.
8. 93·69602912.   9. 1393·7111.

Ex. LVI. (p. 98.)

1. 2·258.   2. 7·0456.   3. 5·9697.   4. 1·0991.   5. (1) 1·17; 204·93. (2) 68·67; ·2803. (3) 72·09544; 5270·76. (4) 4·41958; ·0069993.   6. 20·93.   7. ·095; 19·98.   8. ·613 of it left.
9. $\frac{111}{216}$.   10. (1) 79·8665.   (2) 82·9319.

Ex. LVII. (p. 98.)

1. 1·1375.   2. 16·2945.   3. 81·20812.   4. 3·333715.
5. 4246·48449.   6. 667·81; 114·364272; 3752; 356·40164745.
7. ·01778479; 488·745015235; ·000642.   8. (1) ·9150625.
(2) ·3689.   9. 278·1975 yds.   10. 346⅔ loaves.

Ex. LVIII. (p. 100.)

1. 12·36.   2. 1·236.   3. ·01236.   4. 123600.   5. 123600000.
6. 1737·1.   7. 17371000.   8. 17371.   9. 173710000.
10. 170·01; 170010.   11. ·00521; 521.   12. ·00003; ·03; ·000000003.   13. 108971·6; 1·089716.   14. ·011; ·00011; 110.   15. 2040000; 204; ·00204.   16. 18030; ·001803.
17. 213·2; ·002132.   18. ·0101.   19. ·0008.   20. 124 days.
21. 85·5 times.   22. ·03054.

Ex. LIX. (p. 101.)

1. 6333; 63·333; ·006.   2. ·031; 3·105; ·003.   3. 6221·584; 62215349·056; 62·215.

Ex. LX. (p. 102.)

1. ·25, ·6; 1·5; 6·2; 7·8; ·625; 5·3.   2. ·1875; 8·9375; ·95; ·96875; 7·925.   3. ·94; 4·056; ·006; ·1584; 84·0029296875.
4. ·5078125; 8·75; 76·234375.   5. 3·9125; 16·36.

Ex. LXI. (p. 104.)

1. ·6̇; ·1̇; ·857142̇; ·583̇; ·73̇.   2. 6·037̇; 7·13̇5̇; 100·159̇0̇; 2·8̇823529411764705̇.   3. 11·13̇5̇; 23·012̇3̇6̇.
4. $\frac{2}{9}$; $\frac{5}{99}$; $\frac{2}{11}$; $\frac{31}{198}$; $\frac{1}{37}$; $\frac{2}{7}$.   5. $\frac{17}{30}$; $\frac{368}{495}$; $\frac{20233}{99990}$;
19$\frac{1}{4}$; 20$\frac{7}{10}$.   6. 6$\frac{199}{2270}$; 15$\frac{3}{13}$.   7. 96·523114.   8. 37·443543.
9. 1·817686.   10. 44·494309.   11. 40·8̇; ·258722....; ·01185̇.   12. 2·5416̇; ·136̇; ·0743̇; 30·833953̇.

Ex. LXII. (p. 105.)

1. 75 cents.   2. $4.37½ cts.   3. 62½ cts.   4. 2 qr., 12 lbs., 8 oz.
5. 3 fur.   6. 3 cwt., 2 qr.   7. £1. 3s. 5¼d.   8. 5½d., ·5 q.
9. 3 lbs., 2 oz., 2 dwt.   10. 20 ac., 3 ro., 28 po.   11. 6 ac., 1 ro., 4 po.   12. £1. 8s.   13. 17 wks., 6 days, 5 hrs., 15 min.
14. 16 dys., 12 hrs., 5 min., 45·6 sec.   15. 15 lbs., 3 oz., 2 drs., 2 grs.   16. £12. 3s. 8¼d. ·048q.   17. 12 ac., 2 ro., 4 po., 20 yds., 7 ft., 122·76 in.   18. 80 lbs., 6 oz., 13·23 grs.
19. 66⅔ cents.   20. 9 shillings.   21. 4 cwt., 3 qr. 11 lb., 10 oz., 10⅔ dr.   22. £34. 3s. 4d.   23. 5 lbs., 11 oz., 10 dwt., 24. 6 c. yds., 6 c. ft.   25. 18 ac., 2$\frac{11}{2}$ ro.   26. £2166. 10s.   27. £5. 8s. 3¾d.   28. 3 ro., 11 po., 9 yds., 1 ft., 72 in.   29. 1s. 5$\frac{21}{31}$d.

Ex. LXIII. (p. 106.)

1. ·3̇.   2. ·2̇5̇.   3. ·1458̇3̇.   4. ·81875.   5. ·5416̇.
6. ·00022095̇.   7. ·2̇2083̇.   8. 48·083̇.   9. ·2̇785493827160̇.
10. ·8̇228571̇4̇.   11. ·5375̇.   12. ·87916̇.   13. 4·90̇.   14. ·15972̇.

Ex. LXIV. (p. 107.)

PAPER I.

2. Seventy thousand three hundred and forty; one hundred and twenty-five millions four thousand three hundred and twenty one; five trillions six hundred and seven billions six hundred and five millions two hundred and thirteen thousand four hundred and three.
3. (1) 54502045294; (2) 99276.   4. (1) 1529981369865;
(2) 3875398$\frac{1377}{4947}$.   5. 1372869823.   6. 777348.

ANSWERS.

PAPER II.

1. 3024.   3. 90 pints.   4. 56 feet; 17 times.   5. (1) $239\frac{1}{4}$;
(2) $22540000.   (3) $91870.42 and $8.06 over.

PAPER III.

1. (1) $2\frac{143}{151}$; (2) $2\frac{87}{143}$.   2. $5000.   3. (1) $3\frac{7}{33}$; (2) $3\frac{11}{48}$.
4. £24. 15s.   5. 58 yards.   6. $\frac{1}{8}$ of the orange.

PAPER IV.

1. 60.   2. 84·875 or $84\frac{7}{8}$.   3. ·01236.   4. 416.27\frac{1}{2}$.
5. 21 on smaller side, 24 on larger side, and 72 lookers on.
6. One side scores seven times as many runs as the other, and therefore that side wins.

PAPER V.

1. 12s. 6d.   2. 275s.   3. $42\frac{7}{13}$.   4. $19.90.   5. $5.92.
6. 48.27\frac{1}{2}$.

Ex. LXV. (p. 112.)

1. 27.   2. $6\frac{2}{3}$.   3. 15.   4. $\frac{1}{8}$.   5. 12·64.   6. 15.   7. $\frac{3}{8}$.
8. ·36.   9. $\frac{3}{4}$.   10. 3·2.

Ex. LXVI. (p. 115.)

1. $48.   2. $18.15.   3. 17.33\frac{1}{3}$.   4. 38 bus., $21\frac{7}{11}$ lbs.
5. 20 bus., $28\frac{4}{7}$ lbs.   6. £82. 2s. 8d.   7. 28 cwt., 3 qr., 14 lbs. 12 oz.   8. 44 cents.   9. 29 cents.   10. $8126·01$\frac{9}{16}$.
11. 21 cwt., 3 qr., 18 lbs., 12 oz.   12. $1638·40$\frac{58}{88}$.   13. $61\frac{1}{4}$.
14. 2 mo.   15. 15s. $9\frac{3}{4}d$.   16. £2675. 8s.   17. 3420 steps.
18. £4754. 10s. $10\frac{1}{2}d$.   19. $5606.75.   20. 20 min.
21. 26 yds. 2 ft.   22. 528 pairs.   23. 171 men.
24. $3\frac{1}{3}$ cts., 57.812\frac{1}{2}$.   25. 4.12\frac{1}{2}$.   26. 135 men.   27. 11 hrs., 38 min.   28. $2234.31.   29. $7\frac{1}{2}d$.   30. 12 days.   31. 5s. 6d.
32. 151.14\frac{11}{16}$.   33. £900.   34. (1) £1000; (2) £960.
35. £8. 14s. $11\frac{1}{2}d$. $\frac{48}{73}q$.   36. 30 days.   37. 1902.56\frac{11}{16}$.
38. 3s. 6d.   39. £90.   40. 104 lbs., $2\frac{2}{3}$ oz.   41. 11 hr., $53\frac{11}{44}'$.
42. $5\frac{5}{11}'$ past 1 o'clock.   43. 35·15625 cents.   44. $8\frac{4}{15}$ days.
45. £4005.   46. £132. 0s. $4\frac{3}{4}d$. $\frac{1}{4}q$.   47. 21 days.   48. $10\frac{1}{2}$ hrs.
49. $9\frac{1}{2}$ mo.   50. 12.30 P. M.   10 mi. from place.   51. $5\frac{40}{137}''$.

Ex. LXVII. (p. 121.)

1. 8 wks.   2. 112 men.   3. 64 days.   4. $307.44.   5. $87500.
6. 174 miles.   7. $202·50.   8. 200 horses.   9. 100 months.
10. 2808 qrs.   11. 39 ac., 1 ro. 20 po.   12. 9 mo.   13. 60 men.
(cwt. = 112 lbs.)   14. 91 men.   15. $2\frac{1}{2}$ days.   16. 45 men.
17. 178 qrs., 4 bus.   18. $1·608.   19. $7.20.   20. 4 days.
21. 2 days.   22. $13\frac{1}{2}$ days.   23. 3 lbs., 11 oz., $7\frac{2}{7}$ drs.
24. 25 horses.   25. 180 men.   26. $2\frac{1}{2}$ ft.

Ex. LXVIII. (p. 124.)

1. \$168.75. 2. \$157.50. 3. \$1592.50. 4. \$1927.20.
5. \$3493.75. 6. \$2396.25. 7. £416. 17*s*. 8. £600.
9. \$4965. 10. £6360. 5*s*. 11. £812. 17*s*. $2\frac{1}{2}d$.
12. £1722. 6*s*. 2*d*. 13. £86663. 1*s*. 9*d*. 14. £155668. 10*s*. $11\frac{1}{4}d$.
15. \$267911.87½. 16. \$715024.80. 17. \$72562.35.
18. \$9611.25. 19. £2764. 11*s*. 3*d*. 20. £14. 1*s*. $9\frac{1}{4}d$.

Ex. LXIX. (p. 126.)

1. \$66.50. 2. \$167. 3. \$1496·71875. 4. £11. 11*s*. $3\frac{1}{4}d$.
5. £9. 18*s*. $3\frac{1}{8}d$. 6. £350. 13*s*. $7\frac{1}{2}d$. 7. \$125468.75.
8. \$84.06¼. 9. \$173. 10. \$98.60. 11. \$477·5475.
12. \$127.57[illegible]. 13. \$9.61 $\frac{5}{48}$. 14. \$15.69[illegible].

Ex. LXX. (p. 128.)

1. \$17.38, \$234.63. 2. \$34.76, \$252.01. 3. \$110·71875,
\$638·12375. 4. \$11.22, \$104.72. 5. \$13·25625, \$89·00625.
6. £17. 12*s*. $5\frac{1}{2}d$. +, £80. 11*s*. 3*d*. 7. \$365·755, \$1441·505.
8. \$310.08, \$994.08. 9. £111. 14*s*. $7\frac{5}{13}d$., £7611. 14*s*. $7\frac{5}{13}d$.
10. £171. 9*s*. 9·94…*d*, £5037. 1*s*. 2·94…*d*. 11. 6 years.
12. 8½. 13. £130. 14. £32; 5 fl., 3 c., 0·078125 m. 15. 4.

Ex. LXXI. (p. 130.)

1. \$115.92, \$915.92. 2. \$192.70, \$934.70. 3. \$341.88,
\$901.88. 4. \$28.78, \$336.78. 5. \$103.61, \$713.61.
6. \$229.25, \$1229.25. 7. (1) £1. 1*s*. $6\frac{1}{2}d$. ·88*q*., (2) £6. 19*s*.
$2\frac{3}{4}$d. ·136*q*.

Ex. LXXII. (p. 133.)

1. \$200. 2. \$800. 3. \$1200. 4. \$209.53+. 5. \$900.
6. £129. 6*s*. 9*d*. 7. £179. 12*s*. $10\frac{5}{8}d$. [illegible]*q*. 8. £456. 9*s*.
$11\frac{3}{4}d$. [illegible]*q*. 9. \$42. 10. \$2100. 11. \$95.23[illegible]. 12. \$99.05[illegible].
13. £3. 4*s*. $6\frac{3}{4}d$. [illegible]*q*. 14. $2\frac{1}{2}d$. [illegible]*q*. 15. 4[illegible]*d*.
16. 5 per cent.

Ex. LXXIII. (p. 137.)

1. \$416.79+. 2. \$780.48[illegible]. 3. \$1524.88. 4. \$37.15[illegible].
5. \$15069. 6. \$1391. 7. (1) £10. 16*s*. 4*d*.; (2) £3. 4*s*. 11[illegible]*d*.
8. 6 per ct. per ann^m. nearly. 9. Bank of Toronto. 10. £25.
11. His income less by £64. 12*s*. 12. \$139 $\frac{8}{23}$. 13. £240000
stock. 14. Loss of income = £45. 10*s*. 15. £52. 10*s*.
16. Increase of income = £135. 5*s*. $11\frac{3}{4}d$. [illegible]*q*.

Ex. LXXIV. (p. 141.)

1. \$5.37½. 2. \$250·2903. 3. 29[illegible] cents. 4. \$1900.
5. (1) £6. 5*s*.; (2) £18. 17*s*. 4¼. [illegible]*q*. 6. (1) \$208; (2) \$13.$\dot{1}\dot{3}$;

(8) \$1.55. 7. (1) 10 per cent.; (2) £9. 1*s*. $9\frac{3}{4}$*d*. $\frac{3}{11}$*q*. 8. \$1.20
9. $66\frac{2}{3}$. 10. 5*s*. 3*d*. 11. 4*s*. $1\frac{3}{4}$*d*. $\frac{1}{5}$*q*. 12. $40\frac{1}{2}\frac{40}{79}$
13. 7*s*. $11\frac{1}{4}$*d*. $\frac{4}{7}$*q*. 14. 18*s*. 4*d*. 15. £63. 12*s*. $8\frac{1}{2}$*d*. $\frac{1}{11}$*q*. 16. \$1.20.

Ex. LXXV. (p. 142.)

1. $90\cdot8\dot{3}$. 2. \$5.58. 3. 83·67. 4. 8·667...yrs. 5. $60\frac{1}{4}$ yrs
6. $29046\cdot81\dot{3}$. 7. £191. 8*s*.

Ex. LXXVI. (p. 145.)

1. (1) 224, 336, 448; (2) $\$40.62\frac{1}{2}$, $\$89.37\frac{1}{2}$, \$130; (3) 66 ac., 1 ro., 15 po., 79 ac., 2 ro., 18 po.; (4) £42. 17*s*. $1\frac{1}{2}$*d*. $\frac{1}{7}$*q*., £28. 11*s*. $5\frac{1}{4}$*d*., £21. 8*s*. $6\frac{3}{4}$*d*. $\frac{5}{7}$*q*., £17. 2*s*. $10\frac{1}{4}$*d*. $\frac{1}{7}$*q*. 2. (1) *A* is to have \$136, *B* \$238, and *C* \$306. (2) £136. 10*s*. 3. 1094·3744 lbs. of oxygen, 969·136 lbs. of carbon, 176·4896 lbs. of hydrogen, (cwt. = 112 lbs.). 4. $33\frac{1}{3}$ per cent. silver. 5. 5 mo. 6. 8 mo. 7. *A* ought to have \$6400, *B* \$840, *C* \$720. 8. *A* ought to have received £700, and *B* £900. 9. *A* should pay \$36, *B* \$18, and *C* \$6. 10. 24 men.

Ex. LXXVII. (p. 149.)

1. 14; 17; 25. 2. 29; 30; 42. 3. 49; 87; 98.
4. 111; 200; 623. 5. 703; 763; 509. 6. 1111; 5343.
7. 7906; 5746; 7008. 8. 13509; 6·9. 9. ·094; 21·103.
10. ·025173; ·00003. 11. 7·1414; ·7141. 12. 2·2583; ·2258.
13. 28·3992; 310·3304. 14. ·577. 15. ·166. 16. 2·175.
17. $\frac{48}{55}$. 18. 2·625. 19. 540*s*.

Ex. LXXVIII. (p. 152.)

1. 12; 20; 18. 2. 42; 75; 92. 3. 97; 103. 4. 512; 4·01.
5. 76·3; ·0587. 6. 5·079; 7420. 7. $\frac{2}{3}$. 8. ·643. 9. 1·560.
10. 1. 11. ·464. 12. ·215. 13. 2·154. 14. ·333.

Ex. LXXIX. (p. 152.)

PAPER I.

1. 2 rem. 117257. 2. (1) 15 tons, 8 cwt., 3 qrs., 17 lbs., 1 oz. (cwt. = 112 lbs.) (2) 1 oz. Avoird. = $\frac{175}{192}$ of 1 oz. Troy.
3. £34. 9 fl. 6 c. $8\cdot21\dot{4}28571\dot{}$ in. 4. 1; 2520. 5. (1) 63.
(2) $\frac{5}{6}$. (3) 4 cwt., 3 qrs., 3 lbs. (cwt. = 112 lbs.) 6. (1) 21060.
(2) ·00002106.

PAPER II.

1. (1) \$90. (2) £281. 7*s*. 5*d*. 2. Gain per cent. = \$25; Loss per cent. = \$20. 3. (1) 30502. (2) 256.
(3) 67 yds., 4 in 4. 118625. 5. 2 × 2 × 3 × 3 × 7 × 7 × 13.
6. 34′. $27\frac{3}{11}$″ past 6 o'clock P. M.

ANSWERS 191

PAPER III.

1. 6.   2. (1) 10625; (2) ·030416; (3) The first;
(4) $\frac{29}{275}$.   3. $520.   4. $37\frac{2}{3}$ square yards.   5. $13\frac{41}{44}$.
6. 22.37\frac{11}{32}$.

PAPER IV.

1. (1) $\frac{119}{108}$; $\frac{47}{48}$; (2) 123 times; $\frac{1}{3}$.   2. £139 15s.
3. (1) £20 9s. 6d., (2) 52½.   4. £82.   5. Turkey,
16s. 6d.; fowl, 2s. 10d.   6. $10\frac{59}{225}$.

PAPER V.

1. (1) 3011404; (2) 971472492; (3) 430709070;
(4) 37834342650; (5) 75732561476.   2. 56c.   3. $126.
4. $288.09.   5. $25.40.   6. 1 gal. 2 qts. 1·02 pts.

PAPER VI.

1. 706·85775 feet.   2. 6412032000000000 cubic feet.
3. 8 ft 1⅛ in.   4. $2025.   5. A, $750; B, $500;
C, $250.   6. 18 days.

PAPER VII.

1. (1) 7117423255950;   (2) 6593445483924;
   (3) 68677245810;   (4) 277980278515538;
   (5) 81697259850030.   2. $7.50; $7.52.
3. $12.96.   4. 5000.   5. $868.57; $1709.   6. $\frac{1}{5}$.

PAPER VIII.

1. (1) 10836151080800;   (2) 31808539707205;
   (3) 69154272;   (4) 34175791448.
2. $8.40.   3. $1.62; $1.08; 60c.   4. $812.
5. 420 cents; 140 cents; 560 cents.   6. 31817108.

PAPER IX.

1. (1) 7832; (2) 196734; (3) 3589853148; (4) 3627482760.
2. 52800 yds.   3. $20 to B, $10 to C.   4. $600;
$900.   5. £2190.   6. $3463.85.

PAPER X.

1. 825.   2. $4.20.   3. $312\frac{15}{35}$.   4. ·004125.
5. $200.   6. 130.   7. $3000.   8. $509.62½.

192   ANSWERS.

PAPER XI.

1. 2, 3, 5, 7.   2. 31116.   3. 4 cents.   4.
5. 8 hrs.   6. $\frac{1}{108}$.   7. $120.   8. 101.85\frac{1}{4}$.

PAPER XII.

1. $501.25.   2. $1258.32.   3. $396.   4. $1149.54.
5. $1764.   6. $33306.25.   7. 1348.19\frac{19}{24}$.   8. $1800.
9. 4.88\frac{8}{9}$.   10. 2615f. 7$\frac{417}{869}$c.

PAPER XIII.

1. (1) 228303111$\frac{47342}{160388}$; (2) 7982.   2. 299$\frac{27537}{138437}$ gals.; 3007 lbs., 11 oz.   3. 3 ft. 4$\frac{3}{4}$ in.   4. 822$\frac{2}{7}$ in.
5. 795·9+ yds.   6. 10003712$\frac{88}{101}$.

PAPER XIV.

1. (1) 186798534370. (2) 2511248800235.   2. 200 lbs. 6 oz. 5 dwt.   3. 43 mls. fur. 23 per. 3 yds. 1 in; 10 hrs. 50 min +.   4. 1133, 1339.   5. 1$\frac{8}{77}$.   6. $\frac{5}{8}$.

PAPER XV.

1. 5817600 inches.   2. 10 per cent.   3. 16$\frac{1330}{2051}$.
4. $27.8+.   5. 108$\frac{3}{4}$.   6. $7920.

Mental Arithmetic.

By *J. A. McLELLAN, M.A., LL.D*, *Inspector of High Schools, Ont.*

PART 1.—FUNDAMENTAL RULES, FRACTIONS, ANALYSIS. PRICE, 30c.

PART II.—PERCENTAGE, RATIO, PROPORTION, &C. PRICE, 45C.

W. D. DIMOCK, A.B., H.M.
Provincial Model School, Nova Scotia.

Dr. McLellan's Mental Arithmetic supplies a want that we should have had supplied in our Schools long ago. Same progress cannot be made in Mathematical work, unless what we call Mental Arithmetic is thoroughly and systematically pursued. A boy who is conversant with the principles of Mental Arithmetic, as given in this little text-book, is worth as a clerk or accountant 50 per cent more than the prodigy who can boast of having "gone" through his written arithmetic half a dozen times.

J. S. DEACON, Principal Ingersoll Model School.

Dr. McLellan's Mental Arithmetic, Part I., is a credit to Canadians, and it supplies a long-felt want. It is just what is wanted for "waking up mind" in the school room. After two weeks use of the book with my class I am convinced that it is much superior to any of the American texts that have been used here both as to the grading of questions and the style of the problems.

J. A. CLARKE, M.A., H.M.H.S., Picton.

Dr. McLellan's Mental Arithmetic contains a great number of useful problems well adapted to develop by regular gradations the thinking powers of the pupil, and to suggest similar examples for the use of the teacher.

D. J. GOGGIN, Head Master Model and Public Schools, Port Hope.

Simple in its arrangement, varied in its types of practical questions and suggestive in its methods, it is the best book of its kind that I have examined.

From THE WESLEYAN, Halifax, Nova Scotia.

The series bids fair to take a good place in scholastic work.

The New Authorized Elementary Grammar.

MILLER'S SWINTON'S LANGUAGE LESSONS.

MILLER'S Swinton's Language Lessons is used exclusively in nearly all the Principal Public and Model Schools of Ontario. Among them are

Ottawa, Hamilton, Whitby, Port Hope, Cobourg, Mitchell, Napanee, Brockville, Lindsay, St. Catharines, Strathroy, Meaford, Uxbridge, Brantford, Windsor, Clinton, St. Thomas, Perth, Seaforth, Listowel, Bracebridge, Belleville.

Adopted by the Protestant Schools of Montreal and Levi College, Quebec, Schools of Winnipeg, Manitoba, and St. John's, New Foundland

Resolution passed unanimously by the Teachers' Association, (North Huron), held at Brussels, May 17, 1878: "Resolved, That the Teachers at this Convention are of opinion that 'MILLER'S Swinton Language Lessons,' by McMillan, is the best introductory work on Grammar for Public School use, since the definitions, classification and general treatment are extremely simple and satisfactory."

In my opinion the best introductory Text book to Mason's Grammar. All pupils who intend to enter a High School or to become students for Teachers' Certificates, would save time by using it.

W. J. CARSON, H. M.,
Model School, London.

The definition's in "Miller's Swinton Language Lessons" are brief, clear and exact, and leave little to be unlearned in after years. The arrangement of the subjects is logical and progressive, and the book admirably helps the judicious teacher in making correct thinkers and ready readers and writers.

B. W. WOOD,
1st A Provincial H., P. S., Trenton Falls.

Be careful to ask for **MILLER'S SWINTON**, as other editions are in the market.

www.ingramcontent.com/pod-product-compliance
Lightning Source LLC
Chambersburg PA
CBHW032131160426
43197CB00008B/601